What Makes Airplanes Fly?

Springer
New York
Berlin
Heidelberg
Barcelona
Budapest
Hong Kong
London
Milan
Paris
Santa Clara
Singapore
Tokyo

Peter P. Wegener

What Makes Airplanes Fly?

History, Science, and Applications of Aerodynamics

With 113 Illustrations

Second Edition

Springer

Peter P. Wegener
Engineering and Applied Science (Emeritus)
Yale University
New Haven, CT 06520, USA

Library of Congress Cataloging-in-Publication Data
Wegener, Peter P., 1917–
What makes airplanes fly?: history, science, and applications of aerodynamics / Peter P. Wegener. — 2nd ed.
p. cm.
Includes bibliographical references and indexes.
ISBN 0-387-94784-1 (hrdcvr : alk. paper)
1. Aerodynamics. I. Title.
TL570.W4 1996
629.132'3—dc20 96-23154

Printed on acid-free paper.

Production managed by Steven Pisano; manufacturing by Jacqui Ashri.
Typeset by KP Company, Brooklyn, NY.
Printed and bound by Braun-Brumfield, Inc., Ann Arbor, MI.
Printed in the United States of America.

9 8 7 6 5 4 3 2 1

ISBN 0-387-94784-1 Springer-Verlag New York Berlin Heidelberg SPIN 10538487

To Annette

Preface

Like the first edition of this book, the second edition is addressed to all who are interested in aerodynamics. In addition to the aerodynamics of flight from low-speed craft to supersonic airliners, the topic is interpreted in a broad sense, including the aerodynamics of automobiles, bird flight, and the motion of diverse objects through air or water. The fundamentals of basic mechanics and fluid mechanics—the physics underlying aerodynamics—and general remarks on the nature of science and engineering are interspersed in the text. The use of mathematics is minimal; it is restricted to elementary algebra, and only a handful of simple but basic equations appear. This is in harmony with the expectations of college students in the humanities and social sciences taking courses outside their field, and of general readers whose interest in flight transcends the discussion of specific airplanes, piloting, airlines, airports, and so on.

A brief history of man's attempts to fly from the early days to the Wright brothers and beyond is given. The selection from the multitude of events over the centuries is of necessity incomplete, but some of the major contributions—in particular those that influenced the Wright brothers—are discussed. An overview of the current status of aeronautics and some thoughts on the future of air traffic are given at the end. These discussions are restricted to civilian air transportation. Books on history, aeronautical engineering, and military aviation are listed in Appendix 4 to facilitate further study.

The general reader may safely skip all appendixes and the footnotes. Even the few equations can be dispensed with, since their meaning is explained in the text. However, college students who use this book in general science or engineering courses directed to liberal arts students will find it useful to hear more about the metric system of measurement and subjects such as dimensional analysis. The latter is an important contribution of engineers to human knowledge that gave us, for example, the Reynolds and Mach numbers. Careful study of this material will enable students to solve the problems found in the study guide. The tables of properties in the book are sufficient to extract the numerical values required for their solution.

I developed the material given here while teaching Yale College students in the liberal arts, and I hope that the book may be used similarly elsewhere as a textbook. The instructor ought to be an engineer or scientist with an interest in the fascinating topic of flight, who in addition believes in a liberal education that cannot ignore science and engineering. About a century ago, man rose for the first time from the surface of the earth in gliders and powered aircraft, machines that must be counted among the greatest engineering feats. Who is not curious to find out how it is possible for huge modern airliners to take off at a steep angle and transport people and goods to any place on earth?

I first considered turning my rough lecture notes into a book in 1986, during a term as a fellow of the Institute for Advanced Study in Berlin. The manuscript was written with the support of a grant to Yale University from the Alfred P. Sloan Foundation. Sam Goldberg, who directed the foundation's New Liberal Arts Program, suggested that I write this text. I am deeply grateful to him for his trust in and continuing support of this venture. My colleagues W. Jack Cunningham and Katepalli R. Sreenivasan read the manuscript and provided important advice.

For the first edition of this book published in 1991, John H. McMasters of the Boeing Commercial Airplane Company, Reinhard Hilbig of Messerschmidt-Bölkow-Blohm, and Richard S. Shevell of Stanford University provided much needed advice and data on aircraft aerodynamics. Werner Nachtigall of the University of Kaiserslautern commented on the biological aspects of bird flight.

The second edition at hand, aside from improvements and extensions, provides updated information. This is primarily found in Chapters 10 and 11 on new and planned aircraft. In addition, my somewhat pessimistic earlier thoughts concerning the future of supersonic flight and romantic visions, etc., are put in the context of recent developments. Again I am grateful for Richard S. Shevell's sage advice. Ronald L. Bengelink, the chief engineer in aerodynamics engineering at the Boeing Commercial Airplane Group, provided interesting insights and data on new commercial aircraft. D. K. Hennecke of the Technical University of Darmstadt (Germany), an expert on aircraft propulsion, offered additional thoughts. These colleagues extended valuable aid, but all possible errors are of course my own responsibility. The physics editorial department of Springer-Verlag New York read the manuscript and provided many suggestions. I am grateful for this sensitive advice.

In addition to changes in the text, the Suggestions for Further Reading have been updated and extended. The Study Guide now has been enlarged. In addition, solutions to a number of the problems are now included.

Finally, my gratitude goes to Stephen Mayer, who edited the manuscript of the first edition as well as the new material in the second edition. His background is in the humanities, and when he was stumped, I knew I had to make changes in the text. Susan Hochgraf prepared the drawings, and Mafalda Stock did the final word processing of the first edition.

Contents

Chapter One

A Dream Comes True: The Wright Brothers and Their Predecessors

It is not extravagant to say that the 17th of December 1903, when the Wright brothers made the first free flight through the air in a power-driven machine, marks the beginning of a new era in the history of the world.

Sir Walter Raleigh (1861–1922)

1.1 The First Flights at Kitty Hawk

In the wind-swept sand dunes of Kill Devil Hills, about four miles south of Kitty Hawk, North Carolina, on the Atlantic coast, an old dream of man became a reality. By the toss of a coin on December 14, 1903, Wilbur Wright (1867–1912) won the first turn at piloting the powered airplane he had built with his brother Orville (1871–1948). The weather was beautiful, but the wind seemed insufficient for the Wright *Flyer I* to start from the 9° slope of the dune. The two skids of the aircraft—no undercarriage with wheels existed at the time—were set on a yoke running on a single track. Five men from the local lifesaving station helped with this work. The engine was warmed up while the flying machine was restrained. When all was ready, the men holding it let go, and, after a short run of 35 to 40 feet on the track, the airplane lifted off the yoke into the air. Despite this initial success, the *Flyer*'s nose turned up too quickly, leading to a

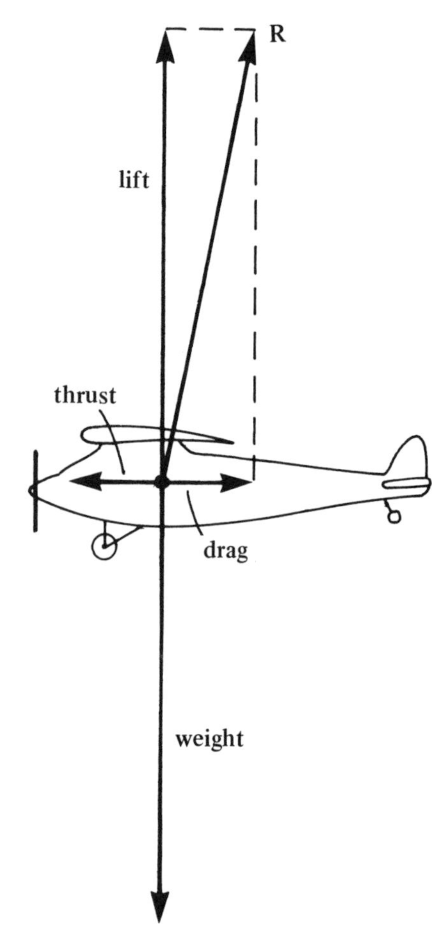

Figure 1.1. The forces acting on an airplane in level flight at constant cruising speed. Lift and weight cancel each other, as do thrust and drag.

stall. A rough landing in the sand broke one of the skids and damaged a control surface, ending further attempts at flight for the day.

What had happened? Stall implies a loss of *lift*. This occurs if the angle of an airplane with respect to the wind direction, the so-called angle of attack, becomes too great. It is the aerodynamic lift force that opposes the *weight* of a heavier-than-air flying machine to keep it aloft. The balance of forces in the vertical direction is shown in the schematic diagram of Figure 1.1, which depicts the fundamental idea of flight. The lift is largely produced by the wings of the aircraft. In cruising flight illustrated in the figure—level flight at constant speed—the lift is exactly equal to the weight of the airplane, and the two forces compensate each other.

On December 17, repairs were completed, and the attempt to fly was resumed. A brisk wind of about 27 miles per hour made the sloping track unnecessary. The wooden rail was now put on smooth, level sand near the building that housed the Wright brothers' workshop. It was a cold day, and the work had to be interrupted by warm-up periods next to an improvised stove made of a carbide drum. Again the five witnesses were present. The Wright brothers, ever methodical, wanted to make sure that their effort, if successful, would be confirmed by independent observers. For years, attempts at flight had been made. Would-be aviators in the United States and in other countries had experimented with airplanes of reasonable or fantastic design. Unfounded claims of success had appeared in several instances, yet no one had in fact accomplished powered flight.

Orville Wright had positioned a camera on a tripod so that the airplane would be in view of the lens when it reached the end of the wooden rail. One of the local witnesses, John T. Daniels, was instructed to pull the string of the shutter at the right moment. Now it was Orville's turn to pilot. Wilbur steadied the right wing, the engine was revved up, and around 10:34 A.M. the Wright *Flyer I* lifted off the rail into a 20- to 22-mph (9- to 10-m/s) wind.* The aircraft wobbled. Gusts of wind, marginal stability, and Orville's understandable inexperience in piloting a powered aircraft combined to produce an erratic flight path. A downward motion forced Orville to land 12 seconds later, after covering a distance of 120 feet, less than the length of a modern airliner. Daniels performed admirably with the camera, producing the famous picture shown in Figure 1.2. Orville later calculated that the airspeed of the aircraft—that is, its actual speed relative to the oncoming wind—was about 45 ft/s or 30 mph (14 m/s or 50 km/h). Had the air been calm, about 540 ft (165 m) would have been covered in the 12 seconds of the first flight. No damage to the machine was discovered, and additional flights were made. The fourth

*For abbreviations and units, consult Appendix 3 and Tables A3.1–A3.2, which include conversion factors between the British and metric systems.

Figure 1.2. *The first flight of a powered aircraft. The Wright brothers'* Flyer I, *with Orville at the controls and Wilbur at the right wing, covered 120 feet in 12 seconds on December 17, 1903.*

and final flight of the day, which took place around noon, now with Wilbur at the controls, covered about 850 ft (260 m) in 59 seconds, foreshadowing the rapid advances that would become characteristic of aviation. After this flight, strong winds turned the machine over, making further flight on that day impossible.

These flights show the remarkable preeminence of the Wright brothers. It was not until November 1907 that a one-minute flight was accomplished in Europe by Henri Farman. The Wright brothers were making thirty-minute flights by 1905. Looking back on this famous day, Charles H. Gibbs-Smith, the foremost historian of early flight, was moved to quote a perceptive British humanist of the turn of the century whose words appear at the head of this chapter. And yet who could have foreseen in 1903 that within one lifetime the modest beginnings at Kitty Hawk would lead to aircraft crossing the Atlantic in three hours at twice the speed of sound, widebody jets carrying hundreds of people halfway around the globe without stopping, and spacecraft reentering the atmosphere in fiery hypersonic flight?

Orville Wright later wrote that "the [first] flight lasted only 12 seconds, but it was nevertheless the first in the history of the world in

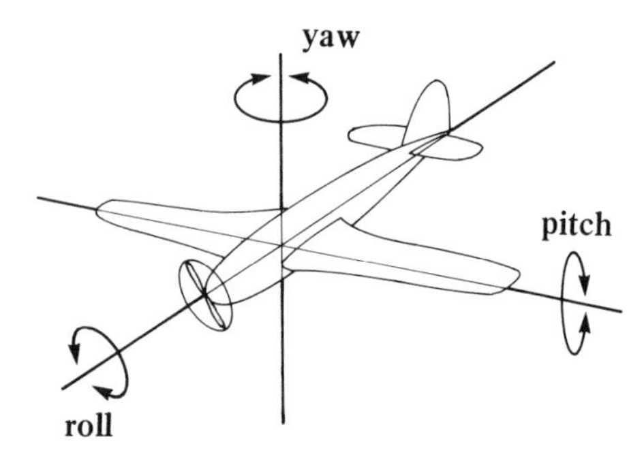

Figure 1.3. The three axes of motion of an airplane.

which a machine carrying a man had raised itself by its own power into the air in full flight, had sailed forward without reduction of speed, and had finally landed at a point as high as that from which it started." He could have gone on to enlarge his definition of flight by adding that the aircraft was *controlled.* In addition to having aerodynamic lift to counter its weight, and *thrust* to overcome the *aerodynamic drag* (Figure 1.1),* an airplane must be controlled about the three axes of motion shown in Figure 1.3. A scheme to achieve control about the *roll* or longitudinal axis of the airplane was the Wright brothers' unique contribution to the development of flying machines.

The two bicycle mechanics from Dayton, Ohio, showed themselves to be brilliant engineers full of original ideas on all aspects of flight. Although they were largely self-taught, we must not view the brothers as merely backwoods tinkerers. Their success was built on careful study of the work of their forerunners and on extensive experimentation, including construction of a wind tunnel, flight tests with tethered and free-flying gliders, and the designing and building of an exceptionally light internal-combustion engine driving two propellers. This epoch-making event merits further discussion, but we will first look at the history of flight leading up to the turn of the century, so that we can see what the world looked like to these gifted brothers.

1.2 Mythology and Legend

To leave the surface of the earth and soar freely high above the valleys and mountains is an ancient aspiration of man. Who among us has not dreamed of flying like a bird? Such primeval fantasies are embedded in the early mythology of peoples in all parts of the world. Thoughts of flight are tied to ideas of divinity. The gods live high above us: the Greek gods occupied Mt. Olympus in northern Greece, and the Christian god dwells in heaven, a vaguely defined place high above the earth and its clouds. Also from the earliest days, the dream of flight has involved the imitation of birds. The Greek god Hermes—a swift messenger—has short wings attached to his sandals and hel-

*In this figure the thrust or push is achieved by a propeller that is driven by a piston engine. The forces of thrust and drag oppose each other. They act in the horizontal direction. In cruising flight these forces are exactly equal, just as we saw that lift and weight exactly compensate each other in the vertical direction. In Figure 1.1 forces are indicated by arrows whose length symbolizes the actual value of the forces and whose orientation shows the direction in which the forces act. Such arrows are called vectors. Viewing Figure 1.1 for cruising flight, you note that no *net* force acts on the airplane. This statement may seem puzzling at first glance, since the airplane flies through the air. The reasons for this fact lie in the laws of mechanics; their application to flight will be given in detail in Chapter 8.

met. Statues of Nike, the goddess of victory, show large wings attached to her back. Similarly, Christian angels are represented from the earliest times with wings.

The best-known mythological aviators in the Western world are Daedalus—an inventor who is sometimes credited with making the first axe—and his impetuous son Icarus. They are shown in Figure 1.4 during their escape from Crete, the Mediterranean island where

Figure 1.4. *The flight of Daedalus and fall of Icarus. Copper engraving of 1690; from Halton Turner,* Astra Castra: Experiments and Adventures in the Atmosphere *(1865).*

they were imprisoned. Daedalus had made wings held together with wax. Icarus flew too high, and when the sun's rays melted his wings, he fell. We will encounter Daedalus again in Chapter 8.4, when we study a more successful latter-day experiment in human-powered flight off the shore of Crete.

We also encounter fliers in the sagas of the Norsemen. Among them we find the Finnish blacksmith Ilmienen—Wieland to the Germanic tribes—who fashioned metal wings to rise from the earth. In the Far East, Shun, the emperor of China around 2000 B.C., was according to legend taught to fly by two princesses at the court, who seem to be the first recorded flight instructors. The aviators in these three ancient tales have in common that they resorted to wings to escape captivity.

Yet other reasons for taking to the air appear in other legends. Chinese writings in the sixteenth century B.C. describe a flying cart. This craft is shown in a later woodcut with wheels that are a cross between paddle wheels and propellers. Early legends in India speak of a flying wooden horse impelled by internal machinery, which could bear a rider. A winged human figure is shown on the grave of Ramses III in Egypt, and the folklore of the African country of Uganda includes flying men. The flying carpet of the Orient and the broom on which witches ride have long ancestries, with the latter possibly using some form of rocket propulsion. All in all, ideas of human flight appear to be as old as spoken or recorded mythology, and many additional examples can be found.

We see a second phase of aeronautics when we look at a few of the many descriptions of flying men not tied to mythology. Such events begin to lead toward serious attempts at flight based on the construction and testing of flying machines.* In the last 2,000 years or so, images of flying men appear in sculpture, paintings, and prints, not to mention absorbing descriptions in books. The magician Simon is said to have flown in a fiery chariot over the city of Rome during the emperor Nero's time. About the year 1020, Elmer, a Benedictine monk of the Abbey of Malmesbury in England, is said to have performed a spectacular glide with wings, starting from the tower of the abbey. He broke his legs in the process. Similar events are described in many countries. For example, the Italian mathema-

---

*The following is based on what is called secondary literature, books and articles (some of which are listed in Appendix 4) dealing with the history of early aviation and its uncertain antecedents and based on study of original documents. Different sources put varying emphases on the relative importance of the contributions of different people, in particular during the last century. The important steps that led to the achievement of the Wright brothers will be emphasized in this account.

tician Batitta Dante reportedly flew a glider across the Lake of Trasimene in 1456. Stories of mechanical models of birds abound: Archytas of Greece described in the fourth century B.C. a model of a pigeon with a pressure mechanism to make it fly. Copper birds that soared and sang are mentioned in Rome around A.D. 500, and an ingeniously devised flying eagle is said to have greeted the emperor Charles V in Nuremberg in 1541. Flying ships appear in the paintings of the famous Dutch artist Hieronymus Bosch (1450?–1516), whose fertile imagination frequently portrayed animals and people in flight.

Fabulous machines filling the skies in increasing numbers can be found up to the present time; imaginary travel by man now extends to interplanetary space and the whole universe. As early as the seventeenth century, Johannes Kepler (1571–1630), who defined and computed the orbits of the planets, wrote a precursor of science fiction. In *Somnium* he told a story about a man who dreams that he is flying to the moon. His report of what he sees there may be considered as the first scientific description of the moon's surface.

Fascination with aviation increased with the rapid development of science and technology after the end of the seventeenth century, and yet romantic notions and fantastic inventions persisted in parallel. An example is shown in Figure 1.5: the great airship of the visionary French writer Jules Verne (1828–1905), who also foretold travel to the moon. In fact, mythology has not disappeared even in our scientific age: UFOs (unidentified flying objects) command much space in the popular press.

1.3 Early Adventures

As a third step in our brief narrative of early flight we single out serious documented attempts to conquer the air—successful or not—that go beyond legend and imagination. Kites were forerunners of airplanes. Like airplanes, they are heavier than air and rise above the ground. They made their first appearance many centuries ago in China and other countries in the Far East, remaining popular ever since. Large kites are said to have been invented in China about 400 B.C. by Mo To Tzu, who constructed kites of light wood that flew tethered—kept leashed to the ground. Closer to our own time, in 1749 the Scottish meteorologist Alexander Wilson, assisted by Thomas Melville, used kites to carry thermometers aloft to study the atmospheric temperature distribution, and Benjamin Franklin experimented with kites in 1752 in his daring and famous studies of lightning. Interesting though it is (and we will return to it later), the flight of tethered kites differs from the free flight that concerns us here.

Figure 1.5. *Vision of an airship by Jules Verne, 1886.*

In describing the Wright brothers' success we mentioned a rigorous definition of sustained flight. However, in pursuing the early history of known attempts to fly, we must abandon this strict notion. In parallel with *ornithopters*—flying machines that emulate the birds by having a man flap wings via some mechanism—and with gliders of various kinds, we admit lighter-than-air craft such as balloons and airships. Indeed, we find that balloons long preceded reliably recorded flights of hang gliders or sailplanes in letting man literally rise from the earth's surface.

One of the earliest sources on the subject is the writings of the scientifically inclined English Franciscan monk Roger Bacon (1214–94?), who understood Archimedes' law of buoyancy (see Chapter 3.2) and suggested that a balloon can float on air just like a ship on water provided it is filled with a lighter substance than air. Bacon did not know that his “ethereal air” could simply be heated air, which is indeed lighter than cold air. Even less could he foresee the discovery of light gases such as hydrogen and helium. But Bacon also mentioned a flying machine with “artificial wings made to beat the air,” providing what is probably the first reference to an ornithopter.

Leonardo da Vinci (1452–1519), the great Florentine painter who in addition excelled in fields ranging from physiology to engineering, provided the first sketches of parachutes and helicopters—a later successor of the spinning top of the Chinese—and for years worked on plans for ornithopters. His astounding aeronautical work came to light with his many other scientific manuscripts only at the end of the last century. Gibbs-Smith reckoned that in these manuscripts over thirty-five thousand words and more than five hundred sketches deal with flying machines, the nature of air, and bird flight. Leonardo had studied the anatomy of both bird and man. Although he was aware of the fact “that the sinews and muscles of a bird are incomparably more powerful than those of man,” he believed that the goal of flight might rouse the extra strength in man to rise in an ornithopter. Gibbs-Smith further suggested that the powerful emotion and symbolism of flight overrode Leonardo’s scientific judgment so clearly demonstrated in other branches of science. Leonardo wrote, “The great bird will make its first flight upon the great swan [Mount Ceceri near Florence], filling the whole world with amazement, and all records with its fame; and it will bring eternal glory to the nest in which it was born.”

Leonardo designed ornithopters for pilots in a prone or upright position, who flapped the wings via pulleys employing both hands and feet, an advance beyond strapping wings to arms only. It is not clear if his beautiful sketches were translated into actual models or even a full-size ornithopter. Other designs show gliders, aircraft that could have been successful in his time.

In the late seventeenth century the Italian Jesuit priest Francesco Lana-Terzi, a professor of mathematics and philosophy, realized that a lighter gas than air is needed to lift a balloon or an an airship, as opposed to a heavier-than-air machine such as Leonardo’s ornithopter. He based this thought on his observation of hot gases rising from volcanic eruptions. Lighter than a light gas would, of course, be no gas at all. Consequently, Lana-Terzi designed an airship to be carried by four evacuated spheres, each with a diameter of 6 m (20 ft). A sail was (mistakenly) added to keep the contraption on course. Vacuum pumps had been built in England and Germany before 1650 at the time of the Thirty Years’ War. Lana-Terzi soon realized that the vacuum sphere’s paper-thin walls could not withstand the external atmospheric pressure but would collapse. He reportedly said that in any case God would not permit the success of such an invention, since it would open up immense possibilities for destruction. Lana-Terzi’s prophecy was wrong, as witnessed by the subsequent successful development of military aviation.

An English contemporary of Lana-Terzi, Robert Hooke (1635–1703), a master of design and experimentation, collaborated at Oxford University with Robert Boyle (1627–91) in discovering a gas law which we shall use in Chapter 4.2 to help us understand the structure of the atmosphere, the medium of flight. Hooke experimented around 1655 with ornithopter models, but he found them "difficult to keep aloft." It is remarkable that such a gifted man did not succeed in making a model fly at a time when science and engineering were coming into full swing.

Although ornithopters have tempted would-be aviators up to this century, their scientific demise as a man-carrying device was pronounced in 1679. The Italian Giovanni Borelli (1608–79) was the first to point out the physiological differences between humans and birds (see Chapter 8.4). He stated in *De Motu Animalium* that man could not hope to support his weight in air "without mechanical assistance." A full understanding of the physiology of humans in relation to flight was arrived at about 120 years later by the father of modern aviation, Sir George Cayley, about whom we will have much to say later. Cayley showed how a powered airplane would be possible by discarding the ornithopter. The following prophetic words define the modern aircraft:

> The idea of attaching wings to the arms of a man is ridiculous enough, as the pectoral muscles of a bird occupy more than two thirds of its whole muscular strength, whereas in man, the muscles that could operate upon the wings thus attached, would probably not exceed one tenth of the whole mass. There is no proof that, weight for weight a man is comparatively weaker than a bird; it is therefore probable, if he can be made to exert his whole strength advantageously upon a light surface similarly proportioned to his weight as that of the wing to the bird, that he would fly like the bird. . . . I feel perfectly confident, however, that this noble art will soon be brought home to man's general convenience, and that we shall be able to transport ourselves and families, and their goods and chattels, more securely by air than by water, and with a velocity of from 20 to 100 miles per hour. To produce this effect it is only necessary to have a first mover, which will generate more power in a given time, in proportion to its weight, than the animal system of muscles.*

Here Cayley correctly predicted the architecture of airplanes propelled by engines, or even by man. However, he could not foresee that about 160 years after he invented the modern aircraft configuration around 1800, the jet age would begin and such machines would

*Charles H. Gibbs-Smith, *Sir George Cayley's Aeronautics 1796–1855* (London: Her Majesty's Stationery Office, 1962), pp. 213–14.

routinely fly with many passengers at speeds of 500 to 600 mph (800 to 960 km/h) halfway around the earth without touching down. In Cayley's time, not even railroads had made their appearance. (The first steam-powered train to haul freight and passengers was the Stockton and Darlington Railway in England, which began operating in 1825.)

We now return to 1709, when a Brazilian Jesuit priest, Father Laurenzo de Gusmaõ, worked with models of paper balloons. He attached a small basket with a fire and proved that a hot-air balloon could fly, long before the Montgolfier brothers captured the imagination of the world. Gusmaõ demonstrated his invention in the palace of the king of Portugal, in the process setting a curtain on fire. In the same year he designed an airship named *Passarola,* which was supposed to be propelled by two magnetic spheres. This was a step backward, but the eighteenth century was ready for ballooning.

The brothers Joseph Michel and Jacques Étienne Montgolfier, of the small town of Annonay near Lyons in France, took up work to build the first successful, large hot-air balloon. Joseph Michel (1740–1810) had studied mathematics, mechanics, and physics, and Jacques Étienne (1745–99) was an architect. As owners of a paper factory inherited from their father, they knew how to handle the task of constructing paper balloons. Experiments with tethered and freely rising models, with the paper skin reinforced by linen, were successful. The final version of their "aerostatic machine," however, was made of linen rendered somewhat flameproof by a chemical. It had a diameter of about 15 m (49 ft), a height of about 23 m (108 ft), and a weight of 1,700 pounds (7,565 N). The huge balloon was filled with 2,200 m^3 (78,000 ft^3) of hot air produced by a straw fire. The fire was lit on the ground below an open basket that hung like a circular balcony below the opening of the balloon.

On September 19, 1783, a ram, a duck, and a cock were loaded into the basket. In the presence of King Louis XVI and 130,000 spectators, the balloon rose freely and remained in the air for seven minutes. Next the Montgolfier brothers enlisted two friends as test pilots. Pilâtre de Rozier—who soon afterward perished as an early victim of aviation—and a major of artillery, the Marquis d'Arlandes, made the first flight of human aviators on November 21, 1783. Starting at the castle La Muette near Versailles, the *Montgolfière,* as it was to be called, drifted about 12 km (7 mi) across Paris in twenty-five minutes, and it landed safely. For the first time man had left the ground for an appreciable time. It took another 120 years to realize the first flight of a heavier-than-air machine, the Wright *Flyer I.*

The heyday of balloon exploration quickly followed. The Academy of Sciences in Paris did not wish to be outdone by inventors in the provinces. The development of a balloon not based on a dangerous open fire had become a clear possibility after the discovery in

1766 of the light gas hydrogen by Henry Cavendish in Cambridge, England. The Academy financed a crash program to support the well-known physicist J.A.C. Charles (1746–1823), who had studied the effect of temperature on the volume of a gas. At the same time the brothers Robert had succeeded in rubberizing silk to make it impervious to leakage of gas. (As you have perhaps already noticed, teams of brothers occur throughout aviation history.)

Charles and the Robert brothers designed a hydrogen-filled balloon, with netting to cover the balloon and support a gondola or basket for the pilots. They added a valve to release gas and provided sand ballast to lighten the load. With these two features, the pilots could descend or ascend at will. On December 1, 1783, only ten days after the first manned flight of the *Montgolfière,* the *Charlière* rose into the air in the Tuileries Gardens of Paris. The first flight attained the astounding altitude of 3,000 m (9,800 ft). Although the Montgolfier brothers remained in the public mind, it was to the more advanced design of Charles that aviators turned in the succeeding years. Charles himself used his balloon with other scientists to explore properties of the atmosphere as a function of altitude; among his experiments he measured the air temperature.

Only quite recently hot-air ballooning has been reawakened as a sport, with the heated air provided by a controlled propane burner and with much use made of advanced lightweight fabrics. We find such colorful balloons joyful things to see, while in Charles's time his balloon appeared like a frightening monster. In one contemporary print a group of peasants demolishes the dragon with pitchforks after it landed in a field.

1.4 From Cayley to the Wright Brothers

It still took over one hundred years after the successful beginnings before lighter-than-air craft became steerable and reliable dirigibles. In turn, at the time of the first balloons the underlying science of flight of heavier-than-air machines was, in principle, at hand. It fell to Sir George Cayley, whom we cited on the futility of the ornithopter, to lay the foundation for modern airplanes. Cayley's vision around 1800 became the basis of all further work in the design of aircraft. As Theodore von Kármán, a leading modern aerodynamicist whom we shall encounter later, said in 1954, "The principle as we know it now, that of the rigid airplane, was first announced by Cayley."

A sketch of this profound invention was engraved by Cayley on the small silver disk reproduced in Figure 1.6. The design for the airplane is shown on the left; it is a biplane (see Figure A2.3). Earlier drawings on paper, which look like the top view of a folded paper

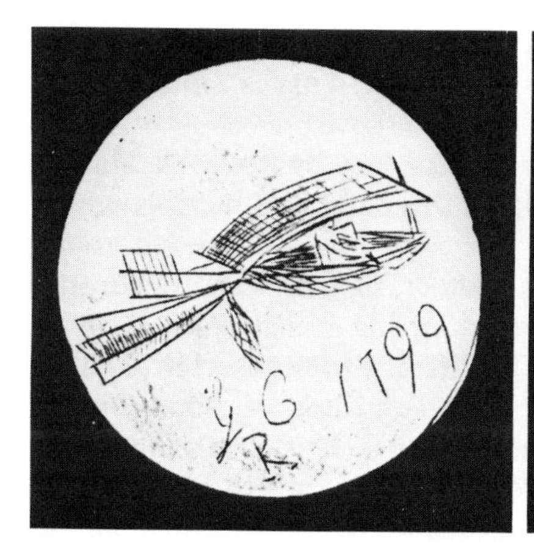

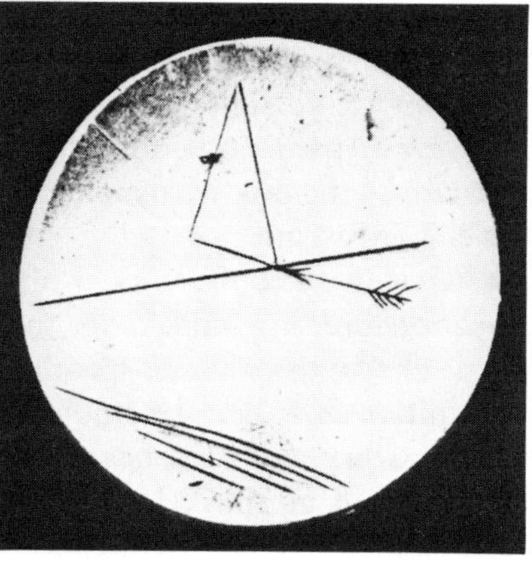

Figure 1.6. Silver disk (obverse and reverse) engraved by Sir George Cayley in 1799, now in the Science Museum in London, showing the first design of a modern aircraft and a force diagram.

airplane, preceded the sketch on the disk. Gibbs-Smith* believed that the curvature of the wing—the camber (see Figure A2.2)—later emulated, for example, by Lilienthal, was the result of experiments with model gliders. There are combined control surfaces of elevator and rudder, and propulsion is provided by paddles at the back of the wing moved by the pilot in the horizontal plane.

This engraving is the first design of an aircraft where *lifting surfaces,* the wings, are separated from the mechanism of *propulsion.* On the right-hand side of Figure 1.6, the back of the disk, we see a diagram of the forces acting on the wings. This is the first time that the resultant air force—R in our Figure 1.1—was separated into the forces of *lift* and *drag* acting respectively at a right angle to and in the direction of the flight path. Cayley advised "to make a surface support a given weight by the application of power to the resistance of air."

As Gibbs-Smith noted, "The propulsion system . . . is of great importance in Cayley's work, and signifies the one great weakness of his life's work in aviation." The paddles shown on the disk follow the many ideas floating around about controlling the flight of balloons. However, propellers must have been known to Cayley, since he had previously used feathered airscrews for his first experiments with a model of a bowstring-operated helicopter. Later in his work he proposed for his models a propeller driven by a stretched rubber band.

*Much of this discussion of Cayley's work is based on Gibbs-Smith's book *Sir George Cayley's Aeronautics 1796–1855.*

The clearest picture of Cayley's visionary airplane is seen in Figure 1.7, a successful model glider. A kitelike wing is mounted on a pole with a weight attached at the front to move the center of gravity forward. Again a cruciform tail unit is fastened by a flexible hinge at the end of the pole. A flyable model of this design is on display at the Science Museum in London.

Cayley was born in 1773 of an aristocratic family at Scarborough in Yorkshire, England. After the death of his father in 1792, he became the sixth baronet, spending his life at Brompton Hall. Following initial schooling, he studied science with tutors and built himself a laboratory in a barn near the manor house. During his long life—Cayley died in 1857, 12 days before his eighty-fourth birthday—he contributed major ideas to several branches of science and engineering besides aviation. Moreover, his technical work was interspersed with activities as a member of Parliament and as a founder of and participant in learned societies.

His work in literally all branches of aerodynamics included an astounding variety of theories, designs, and experiments in addition to those previously mentioned. He tested airfoils—sections of airplane wings—with a whirling arm (see Appendix 2). This testing method was an extension of a 1746 invention of the British engineer and artillerist Benjamin Robins (1707–51), whose work foreshadowed the age of supersonic flight. In fact, Robins measured the drag of spheres flying at supersonic speeds—that is, at velocities exceeding the speed of sound.

Cayley's efforts encompassed ideas on bird propulsion, aerodynamically stabilized projectiles, a hot-air engine, a gunpowder engine, studies of airfoil geometry, ornithopters, undercarriages of airplanes, beam construction, streamlined shapes of minimum air drag, airships, kites, control surfaces, governable parachutes, and other aspects of aeronautics. This catalogue reads like that of an extended contemporary course of study in aviation.

It is not impossible that Cayley actually saw the flight of a man-carrying airplane, the end result of years of study, called the "new flyer" of 1853. He had, after all, built his first full-size glider in 1809. According to the research of Gibbs-Smith, sometime after June 1853 Cayley enlisted his coachman (or perhaps somebody else) as

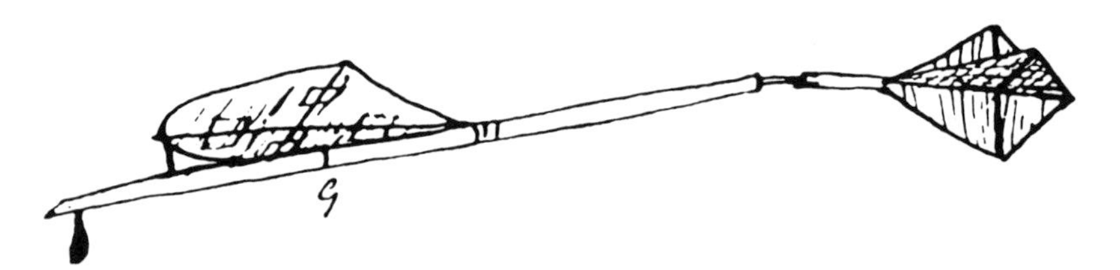

Figure 1.7. Cayley's model glider of 1804.

pilot for a flight in an airplane. Cayley's granddaughter Dora Thompson wrote in 1921, at the age of about eighty, "I remember in later times hearing of a large machine being started on the high side of the valley behind Brompton Hall . . . and the coachman being sent up in it, and it flew across the valley, about 500 yards at most, and came down with a smash. What the motive power was I don't know, but I think the coachman was the moving element, and the result was his capsize and the rush of watchers across to his rescue. He struggled up and said, 'Please, Sir George, I wish to give notice, I was hired to drive and not to fly.'" Although the distance of the flight appears exaggerated, other records of the Cayley family discuss "a machine intended to glide from a height to the plain below without mechanical propulsion." An account of that adventure is also given by Thompson. Whatever truth there is in these descriptions, they make for a good story. Still, although the Wright brothers' claim to fame rests secure, at that time and possibly even earlier, with flying machines such as the often-cited "boy carrier" apparently in existence, some flight may have taken place.

All these late designs had rigid wings, rudder, and elevator to control the yaw and pitch axes (see Figure 1.3), and often some mode of propulsion. It is no accident that Cayley is universally regarded as the founder of modern aviation. As Wilbur Wright said in 1909, "About 100 years ago an Englishman, Sir George Cayley, carried the science of flight to a point which it had never reached before and which it scarcely reached again during the last century."

Early ballooning and Cayley's seminal work took place during a turbulent period of history: the French Revolution started not long after the first balloon flights, Napoleon named himself emperor in 1804 and started the conquest of Europe, and the British burned the White House in 1814 in a final attempt to regain their lost colonies, but aeronautical inventors remained undeterred by these upheavals. Fantastic designs were proposed and experiments carried out by men either unaware of or not interested in Cayley's groundwork. For example, Jacob Degen, a gifted Viennese watchmaker, designed a strange aircraft with two umbrellalike, movable wings to be launched from a balloon. News of Degen induced Cayley to speed up the publication of his treatise *On Aerial Navigation* in 1809.

Strange contraptions appeared throughout the nineteenth century. As late as 1874 the Belgian De Groof fell to his death with his ornithopter, which was launched from a balloon, as shown in the dramatic contemporary drawing reproduced in Figure 1.8. This fatal attempt at flying was totally out of touch with the work of many airmen of the time. However, it is characteristic of such events, since the dream of flight was enticing many men not trained in engineering or science. The tailor A. L. Berblinger of the German city of

Ulm on the Danube River was luckier than De Groof. A master of public relations if not of aircraft design, he widely announced in 1811 his projected plan to fly across the Danube. A crowd assembled to see Berblinger make his historic flight with wings attached to his arms. He started from the city wall near the bank of the river and plunged ingloriously into the water without hurting himself.

Aside from these attempts at flying, the nineteenth century saw a number of more successful ventures. Out of this number we will se-

Figure 1.8. *A fatal launch of a semiornithopter from a balloon in 1874.*

Figure 1.9. The restored steam-driven model aircraft of 1844–47 by Henson and Stringfellow, which looks remarkably like a modern airplane.

lect only a few of the most important forward steps.* In 1842, the Englishman W. S. Henson (1805–88) proposed and patented what we must regard as a prophetic monoplane, an airplane with a single wing (see Figure A2.3). Together with John Stringfellow (1799–1883), a dealer in lace, Henson built and tested a model powered by a steam engine. Of course, the steam engine had set the industrial revolution in motion early in the eighteenth century, but it was designed to stay in one place or to power trains. No light, small engines had yet been developed. So John Stringfellow devised a steam engine putting out up to one horsepower (745 W) and weighing roughly ten pounds. Henson and Stringfellow's first powered model is shown in Figure 1.9. It had a scale of 1:7 with respect to the projected full-size proto-

*Since the early history of flight, including its aberrations, is indeed fascinating, we should mention parenthetically some additional contributors who will be encountered by those delving into history books. They include the Austrian Wilhelm Kress, who around 1880 built powered ornithopters and rigid models that flew, and worked with the Frenchman Pénaud (see below). Kress also proposed a flying boat and built successful propellers. The Russian A. F. Mozhaiski made short hops in powered airplanes. The French army major Clément Ader—long regarded in France as the first man to fly a powered airplane—took off briefly. He later claimed to have flown 33 m (108 ft) in an improved version of his steam-powered *Éole,* but this claim was shown to be unfounded. The Swiss Carl Steiger designed a beautiful, remarkably modern aircraft in the 1890s. The names of the German Karl Jatho and the Hungarian Trajan Vuia appear in the historical literature. We could go on at length. In fact, starting with Leonardo da Vinci, one writer lists 52 published plans for flying machines from 1500 to 1800, including proposals by the American inventor Thomas Alva Edison (1847–1913), who gave us the light bulb.

type and a wingspan of 6 m (20 ft), and weighed about twenty-five pounds. Although the model was a failure, we recognize the surprisingly modern configuration, including an undercarriage and two propellers. A disappointed Henson emigrated to the United States, never to take up aviation again. Stringfellow, in contrast, nearly succeeded in getting another small aircraft to fly in 1848.

Lighter-than-air craft were slowly improving as well. Balloons had still not become steerable, but the Frenchman Henri Giffard built the first nearly successful dirigible. An ellipsoidal balloon covered with netting—remember the design of Charles—and with an open platform suspended beneath was to be propelled by a steam engine. This light engine, which was mounted on the platform, developed 3 HP (2,238 W), driving a large propeller with three blades. The airship had a length of 44 m (144 ft), and it could move in perfectly still air at a speed of 8 km/h (5 mph). On September 24, 1852, this craft lifted off from Paris and flew with a single pilot about 25 km (16 mi) at an altitude of 1,500 m (5,000 ft). The funnel for the steam boiler was directed downward to avoid incinerating the balloon, giving the craft a somewhat bizarre appearance. This event was the first flight by man in a powered machine that could be directed at will, although it required calm air.

Back to airplanes. At the time of Cayley's death in 1857, the Englishman F. H. Wenham was theorizing about flight and the Frenchman F. Du Temple was working on practical problems of aviation, making a hop in a powered aircraft in 1874. Cayley was not known to either one of them. In 1858, a French inventor named P. Jullien was the first to fly a small model airplane with two propellers operated by Cayley's scheme of a rubber band twisted around a spool. However, the first truly successful, stable flying models were designed and flown by another Frenchman, Alphonse Pénaud, in 1871. Pénaud also was unaware of Cayley's legacy, independently advancing ideas on stability prior to the work of the aviators at the turn of the century. His "planophone" shown in Figure 1.10 possessed longitudinal and lateral stability, and it used a twisted rubber band to drive a single pusher propeller, not unlike current toy airplanes.

Meanwhile, aviation was increasingly occupying the imagination of the public. The Aeronautical Society of Great Britain was founded in 1866, and the first airshow ever, a static display without flight demonstrations, took place in 1868 at the Crystal Palace in London, an enormous glass and ironwork building erected for the 1851 Exhibition organized by Prince Albert, the consort of Queen Victoria. At the 1851 Exhibition nothing relating to flight had been shown in the sections of science and machinery. At the 1868 airshow, John Stringfellow showed a triplane model (see Figure A2.3) based on Cayley's ideas, which later influenced Chanute and the Wright brothers.

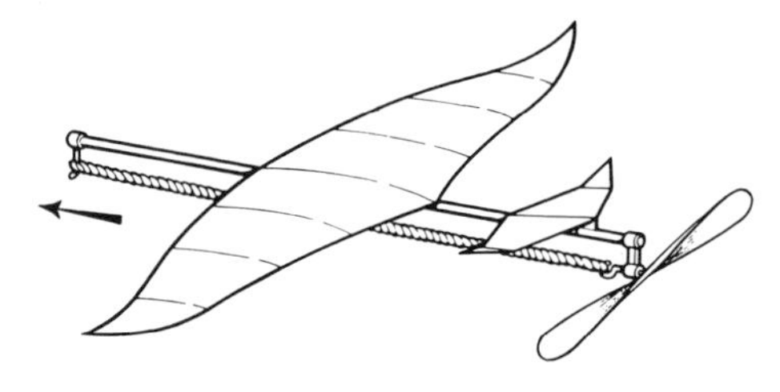

Figure 1.10. *Pénaud's stable, rubber-band-powered model airplane of 1871.*

Further steps taken by the Englishmen Butler and Edwards in 1867 included patented plans for rocket-driven propellers, like those of water sprinklers and pure jets. Their craft had triangular or delta wings, like the space shuttle. In 1879, the Frenchman Tatin flew a model with twin air screws driven by a compressed-air engine. In 1884, the Englishman Phillips patented airfoils (or wing sections) demonstrating that properly curved surfaces—that is, surfaces with camber—dramatically improve the lift. Sir Hiram Maxim (1840–1916) built an enormous steam-driven airplane with a wing span of 31 m (102 ft) and a crew of three. This machine lifted itself very briefly off its test track in 1894. Sir Hiram had moved from Maine to England, and he is best known for his invention of the Maxim machine gun in 1884.

We are now getting close to the end of the century, and it is time to consider one of the great pioneers of flight, Otto Lilienthal, who, according to Gibbs-Smith, "could view the scene of aviation with a shrewd, imaginative, yet realistic mind; who could view it fully in the round, as Cayley had done." This tribute to Lilienthal would have been endorsed by the Wright brothers, who considered him as their primary antecedent.

Lilienthal (1848–96) was born in the small town of Anklam in Pomerania (Germany), where his father was a successful businessman. As a teenager Lilienthal built wings, observed the flight of storks—huge birds native to his area—and developed a consuming interest in aviation. After studies at what is now the Technical University of Berlin, he founded a machine and boiler factory. He was helped in his projects by his brother Gustav, who as late as 1925 published a book on the biotechnology of flight, adding to a famous volume on bird flight (1889) by Otto.

Lilienthal is universally recognized as the first flying human. He developed gliders of the type we would now call hang gliders, as shown in Figure 1.11. Made of wood and canvas with a central frame for the pilot, Lilienthal's glider weighed about forty-five pounds and had a wingspan of 6.7 m (22 ft) and a wing area of 14 m^2 (151 ft^2). About eight of these gliders were made and sold for five hundred German marks each to aspiring fliers in countries as far away as Russia and the United States. Lilienthal was well aware of previous work in aviation. Among other ideas, he adopted camber: his wings were curved on top, departing from a flat plate by one-twelfth of a chord (the distance between the leading and trailing edges of a wing). This shaping gave Lilienthal's gliders a greatly increased lift. In 1895, Lilienthal added biplane gliders to his production. An example is shown in Figure 1.12. It was a configuration that the Wright brothers were to adopt for their airplane.

Lilienthal flew from a number of hills in and around Berlin, including an artificial mound. The top of this mound consisted of an earth-covered hangar. Lilienthal was keenly aware of problems of stability in flight. Proper weight distribution of the flying machine assured stability in the pitch and yaw planes (see Figure 1.3). Lilienthal also considered stability about the roll axis, in contrast to most other aviators preceding him. He achieved roll control in his gliders by shifting his weight, swinging his body from right to left.

Lilienthal made something like two thousand gliding flights in five years. Some of those flights covered distances of up to 300 m (984 ft), and Lilienthal spent a total of about five hours in the air. His dream was to soar, to rise above his starting point, a goal that he could not realize with the materials at hand.

In Lilienthal's time the first steps in the development of the internal-combustion engine were taken. Powered flight was slowly emerging as a reality. In 1876, N. A. Otto constructed the first four-cycle internal-combustion engine. Gottlieb Daimler much improved this machine to lower the weight-to-power ratio, a scheme to squeeze more power out of each unit of weight. Daimler succeeded in gaining a factor of ten with respect to the early models, and Lilienthal began to think about combining engine and glider. In retrospect, he might well have been successful if his work had not come to an abrupt, unexpected, and tragic end. On August 9, 1896, Lilienthal fell with his monoplane glider from an altitude of 10 to 15 m (33 to 49 ft). He

Figure 1.11. Otto Lilienthal on a monoplane glider in 1893.

Figure 1.12. Otto Lilienthal on a biplane glider in 1895.

suffered a fatal spinal injury, dying the day after at the age of only forty-eight years. There has been much speculation about the cause of the accident; it is not impossible that Lilienthal achieved for a moment soaring motion rather than gliding flight, causing a momentary lapse of attention. Stall—the loss of lift—may have ensued, and he crashed.

It is odd that Lilienthal had no followers in his native country. In fact, the German contribution to the development of early aviation after this great man was minor until the years shortly before the First World War. Lilienthal's brother did not succeed in carrying on the work, and he emigrated to Australia a disappointed man. Fortunately, the tradition was carried on elsewhere. A Scottish student of Lilienthal's, Percy S. Pilcher (1866–99), who was a lecturer at the University of Glasgow, in 1895 built a monoplane glider, the *Bat,* that displayed better stability than his master's flying machines. Pilcher's work culminated in his fourth and last glider, the *Hawk,* a larger and slightly heavier craft. He added an undercarriage to soften the first bounce in landing and improved the stability (he understood the need to control the roll axis). He even thought of installing an engine. He flew 250 m (820 ft) across a valley with the *Hawk*. But again a successful career ended prematurely in disaster. During a flight in September 1899 near Rugby, the tail of his glider broke, and Pilcher died like his teacher.*

In viewing the history of flight, Gibbs-Smith separated prospective pilots into "airmen" and "chauffeurs." The airmen viewed the three dimensions of space open to flight as their territory. Consequently, they worried about control of the airplane about all three axes (see Figure 1.3). Cayley and Lilienthal are outstanding examples of airmen. The chauffeurs, in contrast, thought of flight as a powerful extension of surface transportation; all that was needed was to push a machine up into the air. Maxim, Ader, and others represent this group. Only airmen recognized the unique difference of the new medium to be conquered; the science of flight and the art and skill of piloting had to be mastered prior to a successful powered flight that would meet the rigid definition of success outlined in the description of the Wright brothers' first flight at the beginning of this chapter.

Octave Chanute (1832–1910), a French railroad engineer who later emigrated to the United States, became fascinated with aviation at the relatively advanced age of sixty, shortly before Lilienthal's death. He collected much information about aeronautics, publishing a book called *Progress in Flying Machines* in 1894. The Wright brothers read this text and became his friend. A. M. Herring (1865–1926), an American follower of Lilienthal, joined Chanute as his co-worker and pilot. They owned and flew a Lilienthal glider, and they designed and built their own gliders, which they flew on the shores of Lake Michigan. Herring added a compressed-air engine and a propeller to his hang glider. In 1898 he succeeded in performing short flights barely above the ground. The Wright brothers were much influenced by the Chanute biplane design shown in Figure 1.13. Like Chanute, they became admirers of Lilienthal, and thus a close succession emerged. The stage was now indeed set for the addition of an engine to an aircraft.

But we must make one last detour before returning to the starting point of this chapter, the first powered flights at Kitty Hawk. This detour concerns the Wright brothers' American competitor Samuel Pierpont Langley (1834–1906). After graduating from high school in Roxbury, Massachusetts, Langley studied astronomy and engineering on his own at the Boston libraries. In 1866 he became a professor of physics at the Western University of Pennsylvania (now the University of Pittsburgh). His major contributions to astronomy include measurements of the spectral energy distribution of sunlight, the invention of an instrument call the bolometer to make the measure-

---

*Philip Jarrett, *Percy Pilcher and the Quest for Flight* (Washington, D.C.: Smithsonian Institution Press, 1987).

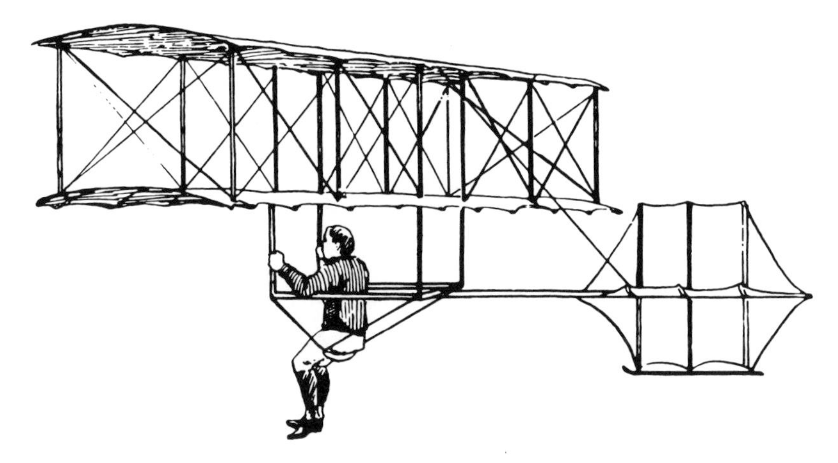

Figure 1.13. *Octave Chanute's biplane glider of 1896.*

ments, and studies of solar and lunar spectra. Based on his scientific achievements, Langley became the head of the Smithsonian Institution in Washington, D.C., in 1887. There he established an astrophysical observatory and the National Zoo.

Aviation caught Langley's attention in 1886, after he heard a lecture on flight at the annual meeting of the American Association for the Advancement of Science, a learned society that still flourishes. As a result, he turned his interest to powered model airplanes and the design of light engines. He built more than a hundred models with rubber-band motors, but he was not happy with any of them. Next he constructed larger steam-powered models with wingspans up to 14 ft (4.3 m) and secured financial support for further work from the Smithsonian Institution. In 1896, the year of Lilienthal's death, he demonstrated remarkable flights of up to 4,200 ft (1.3 km) with two of his models.

These accomplishments made Langley famous, and in 1898 he obtained a grant of $50,000 from the War Department to develop a full-scale manned airplane. Charles M. Manly, Langley's chief assistant and pilot, was an ingenious designer of gasoline engines. He constructed a light radial engine—the cylinders were arranged in a star pattern—that developed 52 HP (39 kW) to spin two propellers at 575 rpm (revolutions per minute). Langley's huge aircraft, which had tandem wings of 50-ft (15-m) span, was strangely named the *Aerodrome.* It was to be launched by a catapult from the roof of a houseboat moored in the Potomac River in Washington, as shown in Figure 1.14a.

Unfortunately, Langley was no airman. His ideas of stability and control lagged far behind the knowledge of the times. Manly had no prior experience with gliders; he was really just going along for the

a

b

Figure 1.14. *(a) On October 7, 1903, Samuel Langley's* Aerodrome *piloted by Charles Manly is ready for launching on a houseboat in the Potomac River in Washington, D.C. (b) After the launch.*

ride. The craft was launched on October 7, 1903, and it fell straight into the Potomac, as seen in Figure 1.14b. The launching mechanism was blamed for the initial setback. Fortunately Manly was not hurt. The hapless pilot again entered the cockpit on December 8—recall that the Wright brothers flew successfully only nine days later—and the second start ended even more disastrously. This time the aircraft not only fell as before, but it disintegrated into a tangle of parts. Again Manly was lucky. After being fished out of the Potomac, this gentle man is reported to have broken out in a "most voluble series of blasphemies."

Langley was attacked in Congress and the press, and the last years of his life were overshadowed by his failure. This is a sad fate in view of his contributions to science. However, to this day Langley has supporters who claim that the launching method of the *Aerodrome* was at the heart of the failure and that—just like the successful models—it could in principle have flown. Attempts were made in 1914 to demonstrate that Langley's refurbished machine could indeed fly with an improved wing structure. Equipped with floats, the aircraft lifted off a lake in New York State to fly for about 150 ft (46 m). Glenn Curtiss, about whom we will be hearing more later, was at the controls. This gifted airplane designer had had a hand in making important changes to the craft which made it greatly superior to the original 1903 *Aerodrome.*

The 1915 Annual Report of the Smithsonian Institution contained the claim "that former secretary Langley had succeeded in building the first aeroplane capable of sustained free flight with a man." This obviously groundless claim led to continued disagreement with the Wright brothers.* In fact, it was not until 1948 that the Smithsonian accepted the original Wright *Flyer I* to be exhibited next to Lindbergh's *Spirit of St. Louis,* which had already entered that shrine of aviation. The Wright *Flyer* had previously been housed in the Science Museum in London. The acceptance of the priority of the Wright brothers was finally official.

Langley's name, however, is permanently and properly attached to the Langley Laboratory of the National Aeronautics and Space Administration (NASA) in Virginia. Langley's pilot Manly—who had a later career in the aviation industry—should also be remembered as the innovator who designed the radial internal-combustion engine that was to play an important role owing to its low air drag (see Chapter 9.2).†

---

*The story shows the foresight of the Wright brothers in arranging for proof of their flights prior to execution, since they preferred to work far from premature publicity. Still, when their first flights were announced after the fact, the *Dayton Journal's* reporter Frank Tunison remarked that if "the flight would have been 57 minutes rather than seconds, it might be newsworthy." For an account of the disagreement, see Tom D. Crouch, "The Feud between the Wright Brothers and the Smithsonian," *American Heritage of Invention and Technology* (1987).

†The complex technical and human relations among the early American pioneers of flight are well described by Tom D. Crouch in *A Dream of Wings,* which is cited in Appendix 4.

One of the prime elements in the success of the Wright brothers at Kitty Hawk was their control of the airplane about all three axes of flight (see Figure 1.3). The pilot in the prone position (see Figure 1.2), which was chosen to reduce the air drag, operated a lever that controlled the forward elevator. In addition, movable cables were attached to the outer edges of the upper wing of the biplane, and fixed cables connected the upper wing with the lower wing. The movable cables were linked to the rear rudder and a cradle that moved at a right angle to the longitudinal axis of the aircraft. The pilot's hip rested in the cradle. In order to circle to the left, the pilot slid his body—and the cradle—to the left. This motion turned the tail rudder to the left and imparted an increased angle to the right wing. The combined effect lowered the left wing and turned the craft to the left. Ideas about how to control roll, culminating in the ingenious wing-warping technique of the Wright brothers, had been proposed by several aviators. The French inventor L. P. Mouillard had published a book called *L'Empire de l'Air (The Empire of the Air),* in which he collected all these observations on flight, addressing in particular the stability problem. This treatise was known to Chanute, who had passed it on to the Wright brothers. Their scheme proved to be successful in 1903; their later aircraft separated the control functions. The Wright brothers patented their schemes of airplane control. However, a series of lawsuits concerning priority of ideas caused them many problems.

Both the 1903 and later aircraft of the Wright brothers were in fact inherently unstable. Stability implies that if the pilot lets go of all controls, an airplane stays on its course; it should return to stable flight after being buffeted by gusts or other destabilizing effects. More of this will be discussed in Chapter 9.1. The Wright brothers were not afraid of an unstable airplane, one whose controls must continually be handled to compensate directional changes in order to stay on a straight path. The bicycle is also an unstable device without a controlling rider, and making bicycles was of course the Wright brothers' original profession. Unstable flight, as we shall see later, has the advantage of requiring only small forces to move the control surfaces, but it also requires first-rate piloting. This mode of flight was changed by later aviators, but the handling of the Wright *Flyers* was an integral part of their success, since the force needed to operate the controls had to be exerted constantly by the pilot himself.

In 1904 the Wright *Flyer II* made about eighty flights, including turns and circles, and in 1905 it banked, turned, and flew in circles and figures of eight. It exceeded in two instances a flight time of half an hour and thus deserves to be called the world's first practical airplane. In 1908 the Wright brothers toured Europe demonstrating their machine, which was far advanced in comparison with all local products. They were eminently successful with their demonstrations and influenced all future aviators. The French had been dominating flying in Europe. Henri Farman flew one minute in 1907 and attained two minutes in the air in 1908. Meanwhile the Wright brothers had flown with passengers, negotiated with the U.S. Army, and produced aircraft for sale. After supporting their research and testing for many years, they needed commercial business. Their new airplane, the Wright *A*, which they brought to France, flew one hour and thirty-one minutes in September 1908 and two hours and twenty minutes in December, piloted by Wilbur. This was the year that Orville crashed after a propeller broke. He was injured, and his passenger, Lt. Thomas E. Selfridge, died in the accident.

On October 4, 1909, Wilbur again flew his plane—with a canoe attached—on a 21-mile (34-km) round trip between Governor's Island in New York harbor and Grant's Tomb. In 1909, the army finally bought the Wright brothers' airplane. Wilbur died tragically of typhoid fever in 1912, but Orville lived a busy life until 1948, contributing to aviation to the end. In fact, there exists a picture of him smiling while piloting a Lockheed Constellation aircraft. The Wright brothers' achievement is nicely summarized by Malcolm Ritchie:* "While Manly was being fished from the water, the only two trained pilots in the world were at Kitty Hawk just two weeks from their historic flight. They had gotten there by becoming competent and methodical engineers, patient and accurate scientists, and finally test pilots. To do all this in 4½ years deserves to be ranked among the greatest intellectual achievements in technical history. It is likely that their achievement will be more highly regarded as time goes on. That may be true of Lilienthal, also."

Spurred on by the Wright brothers, French aviation now began to flourish. The first flying competition ever was held at Reims in 1909. Henri Farman flew over three hours in an airplane designed by Gabriel and Charles Voisin (brothers again). Glenn Curtiss, the Langley supporter whom we mentioned previously, established speed and altitude records. Thirty-eight flying machines were entered at the Reims air meet. Of these, twenty-three actually flew, and eighty-seven flights covered more than 5 km (3 mi). The French excelled at monoplane or single-wing airplanes, with a tail in the rear carrying rudder and elevator. Various schemes were employed to control the roll axis. Much of the success enjoyed by French planes was based on an aircraft engine built by Leon Levavasseur called the Antoinette. This popular engine had eight cylinders producing 50 HP (37 kW).

---

*Malcolm L. Ritchie, "The Research and Development Methods of Wilbur and Orville Wright," *Astronautics and Aeronautics* 16(7/8):56 (1978).

Figure 1.15. Louis Blériot crosses the English Channel in his monoplane on July 25, 1909.

The aeronautical event of the year 1909 was the first successful crossing of the English Channel by Louis Blériot (1872-1936) on July 25, shown in Figure 1.15. His aircraft, the Blériot XI, which was constructed starting in 1908, was powered by a 25-HP (19-kW) engine. Blériot's success led to sales of his monoplane all over the world.

The early days of flight were closing. Airplanes were now being built in many countries, and soon World War I would bring even more rapid advances. But before closing our survey of the early period of the pioneers, we return to a remark by Cayley. He had predicted that a hundred men would die before powered flight was a reality. Fortunately, for once he was wrong. From 1896, the year of Lilienthal's death, to the end of 1909, when Blériot crossed the English Channel, only seven men lost their lives in accidents involving gliders and powered airplanes. Cayley's one hundred deaths did occur by the end of 1910, if pilots and passengers are counted. Jumping ahead to our time, we find that flying is very much safer per mile traveled than riding in an automobile. Finally, when we note that many persons that were alive at the time of the first flight are still alive today, we realize that we have been witnessing a technological development of breathtaking rapidity.

Chapter Two

Milestones of the Modern Age

We are progressing rapidly in the construction of the monoplane and prospects for a successful flight appear brighter as time goes on and information comes in.

Charles Lindbergh (1902–74)

2.1 Notes on Aeronautical Research

We are now faced with a dilemma. In the period following World War I—and in part stimulated by developments in military aviation—there was a rapid advance of research in aerodynamics, design and construction of aircraft, development of engines, and all other aspects of aviation. The last chapter presented a selective survey of the early days of flight up to the beginning of this century. In this chapter, however, it will be impossible to trace the further history of flight, a field that must be left to the extensive literature that is readily available. In particular, we must leave the history of the development of airliners and air traffic in all its aspects, including mail, freight, and passenger transportation. Progress in aviation took place in many countries, with the United States frequently in the lead. The current status of aviation, including some data on aircraft, will be discussed briefly in Chapters 10 and 11. Now we shall concentrate on a few special, even dramatic, events that caught the imagination of many.

But first, we should consider briefly the way the U.S. government has supported research and development in this field over the years. By 1914, the year of the outbreak of World War I, the European nations had all developed aeronautical capabilities to varying degrees. The war sped up the progress of aircraft and dirigible (airship) technology. To advance U.S. interests, and not to be left behind, President Woodrow Wilson, on March 3, 1915, founded the National Advisory Committee for Aeronautics (NACA), a body that consisted of scientists and engineers with a strong interest in aviation. The committee met regularly to advise the government and to see if some order could be brought to the diverse efforts to design and build aircraft.

In 1915, NACA suggested new initiatives, proposing that the government itself build laboratories for research and development to keep this country competitive in the field. In 1917, the United States entered the war in Europe. Research in aeronautics had now become more urgent, and it was decided to follow NACA's advice and build a laboratory primarily devoted to aeronautical research. This facility was opened at Langley Field, in Virginia, at the site of an existing government research station named after Samuel Langley, as we saw in Chapter 1. Several additional specialized laboratories followed in due time. The results of the research and development in the many wind tunnels, test stands, and other facilities at the NACA laboratories were communicated to airplane manufacturers and the public by a series of reports that continues to this day. This work provided the major impetus to the later successes of aviation in the United States.

One typical example of the outstanding work done by NACA is the systematic investigation of airfoil or wing characteristics by wind-tunnel experiments and by calculation. In these studies the values of drag and lift, pressure distributions, and the like were recorded for many geometries or shapes, leading to an extensive increase in knowledge. Whole families of airfoils were generated. This was accomplished, for example, by variations of the camber (see Figure A2.2). Catalogues of these results were distributed widely, boosting aircraft design around the world.

Aerodynamics was also studied at university laboratories. In particular the Guggenheim Foundation initiated support for aeronautics at the California Institute of Technology in Pasadena, California. Theodore von Kármán (1881–1963), a Hungarian who in 1908 had received his doctorate at the University of Göttingen under Ludwig Prandtl's direction (more about him later), was invited in 1930 to join Cal Tech. He gave up his position at the Technical University of Aachen in Germany and moved to California. There he attracted many students, and his group spread the word about aerodynamics throughout the country. Modern fluid dynamics and aerodynamics took hold in the United States. A large wind tunnel was built at Cal Tech (see

Figures 6.1, 7.11), in which such famous airplanes as the Douglas DC-3 were developed.*

NACA later also entered rocketry, as did other laboratories such as the Jet Propulsion Laboratory at Cal Tech, in which von Kármán again was influential. Some thought was given to space exploration, but the major push in this direction was dramatically provided by the successful launching of the first Soviet Sputnik on October 4, 1957. The mission of NACA was extended, and the organization was converted into the National Aeronautics and Space Administration (NASA) on October 1, 1958. This new unit of the government became an independent civilian agency under the executive branch of the federal government. In addition to taking over the original goals of NACA, NASA proceeded to work in all aspects of aeronautics and astronautics. On January 31, 1958, the first Explorer research satellite was put into orbit by the United States. The scientific instruments in its payload of 14 kg mass led to the discovery of regions of dense radiation at an altitude of about 1,000 km, now called the Van Allen belt after the experimenter who designed the research. Like this early mission, much of NASA's current work is done in cooperation with universities and other research groups.

The Apollo program put man on the moon, successful planetary probes were launched, the surface of Mars was analyzed, Jupiter and its moons were explored, and the space shuttle was sent into orbit many times. Commercial and military satellites are frequently fired by rockets into orbit. Satellites are deployed—or even retrieved. The big astronomical space telescope *Hubble* was repaired in orbit, a feat that a few years ago would only have appeared in science fiction. These accomplishments have been widely publicized by the press and television; some of the early events we are about to consider are much less well known.

2.2 Great Moments in Aviation

As noted at the beginning of this chapter, we shall recount only a few special events in the history of the rapid advance of flying. The choice is somewhat arbitrary. All of them are significant. Some are famous, while others have been nearly forgotten. A discussion of some interesting events—in particular lesser-known ones—is better for our purposes than a more complete listing. One useful compilation of the

---

*See Paul A. Hanle, *Bringing Aerodynamics to America* (Cambridge: MIT Press, 1982), and Richard P. Hallion, *Legacy of Flight: The Guggenheim Contribution to American Aviation* (Seattle: University of Washington Press, 1977).

history of flight* gives a chronological listing of all significant events from the earliest time to the present. This encyclopedic catalogue can be used to find more details on the material of this chapter, and it makes us aware of how selective we had to be. Additional material can be found in various other books listed in Appendix 4.

Alexander Graham Bell (1847–1922), the inventor of the telephone, became interested in aviation through the influence of Samuel Langley. Bell invented the tetrahedral kite and thought of building airplanes. In 1907, he got together with four like-minded men, including Lt. Thomas Selfridge, who was to perish in the 1908 crash of Orville Wright's plane, and Glenn H. Curtiss (1878–1930), who would excel in the 1909 Reims air meet. Bell had met Curtiss—who held a motorcycle speed record, among other accomplishments—at a show where Curtiss exhibited one of his aircraft engines. Two Canadian engineers completed the group, which was named the Aerial Experiment Association.

The association built a series of aircraft with mixed success. In 1908 it entered a distance contest sponsored by the magazine *Scientific American*. Glenn Curtiss piloted the *June Bug* (all its craft were named after insects) for a distance of 5,090 ft (1.55 km), exceeding the required one kilometer. This event was proclaimed as the first public flight, despite the fact that the Wright brothers had made flights of up to 24 miles (39 km) outside Dayton, Ohio, in 1905. Again we note the Wright brothers' unwillingness to seek publicity.

The breakthrough for Curtiss came in 1910, when he flew from Albany to New York City in an advanced version of the Reims aircraft, the *Golden Flyer*, powered by a 50-HP (37-kW) V-8 water-cooled engine of his own design. This aircraft followed the Wrights' pattern of having the elevator forward and the rudder behind the wings. With this achievement, Curtiss won a $10,000 award offered by the *New York World*, and he was received like a hero in New York City. His fame as a daring aviator and designer was firmly established in the public mind.

In 1911 Curtiss tried to obtain a patent for his design of ailerons (see Figure A2.1). This led to an unfortunate patent fight with the Wright brothers, while ultimately a third aviator, Dr. William W. Christmas of Washington, D.C., received this important patent. As we have seen, the basic aerodynamic idea of roll control by wing-warping originated with the Wright brothers. Curtiss established in 1909 the first flying school in the United States, where he trained many pilots, and he set up his own company and produced the eminently successful Jenny biplanes during World War I.

---

*See Michael J.H. Taylor and David Mondey, *Milestones of Flight*, cited in Appendix 4.

Figure 2.1. The Navy Curtiss flying boat that crossed the Atlantic in stages in 1919.

The age of competitions and prizes was in full swing. The publisher William Randolph Hearst established a $50,000 prize for the first coast-to-coast flight. Calbraith P. Rodgers, a successful pilot at a prior air competition in Chicago, attempted this feat in a modified Wright airplane. In 1910 he struggled for forty-nine days, making sixty-nine stops between New York City and Pasadena, California. But he exceeded the set time and did not win the prize.

In 1913, Lord Northcliffe, the publisher of the London *Daily Mail,* announced a prize of £10,000 for the first nonstop crossing of the Atlantic Ocean. The war intervened, but in 1919 the offer was renewed. Glenn Curtiss had designed the largest seaplane to date, the U.S. Navy Curtiss (NC), which was a leading competitor. In the NC, four engines were mounted between biplane wings spanning 126 ft (38 m), and the fully loaded flying machine weighed 28,000 lbs (Figure 2.1). The navy's plan was to have four of these airplanes fly from Newfoundland to the Azores, islands in the Atlantic at the latitude of Spain, although this course did not make it possible to win the prize. The pilots and crews were provided by the navy. Following the long flight from the United States to Newfoundland, NC-2 had to be gutted to provide spare parts for the other three planes. These repairs would take time.

But the navy group was not alone. Many veteran fliers of airplanes and airships, recently released from wartime service, were anxious to prove their daring and abilities by an Atlantic crossing. The scene for their exploits was an appropriate one. Newfoundland juts out toward Europe from the North American continent. Flights between the eastern United States and Europe over Newfoundland follow a curve—a great-circle route—that connects two points if you stretch a string between them on the globe. Current airliners pursue the same course. Provided there are no unusual weather conditions, you make landfall at the barren coast of Labrador or Newfoundland when you fly from Europe. An additional incentive for flying from Newfoundland to Europe is that the prevailing winds blow from west to east, as we shall see in Chapter 4.3 when we discuss the circulation of the atmosphere.

All in all, eight British planes got ready near St. John's, Newfoundland, in addition to the three U.S. flying boats. Two dirigibles, the poorly tested British R-34 and the German zeppelin L-72, also stood by. The latter was the most likely to succeed: it had been designed during World War I to fly without refueling from Germany on a round trip to bomb New York City. However, considering the short time since the war—zeppelins had in fact bombarded London—the Germans decided to scrap this venture. The U.S. Navy had in addition entered its nonrigid airship C-5. The C-5 had no central rigid structure; it was in essence a cigar-shaped balloon with a suspended cabin, engines, and controls. It had completed the 1,440-mile (2,250-km) trip from Long Island without difficulties.

With this armada present in May 1919, the weather turned bad. The mooring lines of the C-5 ripped loose. After frantic, unsuccessful attempts by the ground crew to deflate the ship by ripping the gas line, it took off, without anyone on board, toward the ocean, never to be seen again, a Flying Dutchman of the air.* One British pilot, the much-publicized Harry Hawker, and his navigator Kenneth Mackenzie-Grieve took off in a single-engine Sopwith, an airplane

*See John Toland, *The Great Dirigibles,* cited in Appendix 4.

that had been highly successful during the war. After covering over 1,000 miles (1,600 km), the plane's engine gave up. During the last moments of the flight the unlucky fliers spotted the steamer *Mary*, landed nearby, and were eventually picked up.

The three U.S. Navy seaplanes, delayed by weather and repairs, took off on May 18. Destroyers were positioned 50 miles (80 km) apart to provide navigational beacons or to assist in a rescue. Airplanes and destroyers soon lost contact with each other. However, NC-4 under Commander A. C. Read reached the Azores safely after a remarkable flight of fifteen hours, part of the way in heavy fog. The other two planes lost their orientation, and their pilots decided to come down on the water to determine where they were. Both encountered heavy seas that caused sufficient damage to make further flight impossible. NC-1 was feared lost at sea, but the crew had been picked up safely by the steamship *Ionia,* whose wireless could not make contact to tell of the rescue. NC-3 performed an unusual and dangerous sea voyage, with its partly damaged engines pushing it 200 miles (320 km) to the Azores. A few days later NC-4 reached Lisbon, Portugal, refueled, and arrived at London, its final destination. This was the first crossing of the Atlantic by an airplane.

One of the British airplanes, a Vickers Vimy used during the war as a twin-engine bombing plane, was ready to start on June 14. It was piloted by Captain John Alcock (later Sir John), and the navigator was the American-born Arthur W. Brown. After a difficult flight of sixteen hours and twenty-eight minutes in dense fog and with a dead wireless, the aircraft landed somewhat embarrassingly in an Irish bog. The difficult takeoff of the overloaded plane and its landing are shown in Figure 2.2. Since this feat was the first nonstop crossing of the Atlantic—in contrast to NC-4's adventure—it won the prize. Commenting on this event, Toland suggests that "the first crossing was made in a plane that no one in his right mind would now fly from Newark Airport to La Guardia Field." Taking a look at this small and rather flimsy plane, which is now on display at the Science Museum in London, and comparing it with Curtiss's seaplane, one is tempted to agree with this remark.

A most unexpected and truly dramatic flight was quickly to follow. The British dirigible R-34, under the command of Major G. H. Scott, left Scotland on July 2 for North America. On board as an observer was General E. M. Maitland, the head of the British balloon forces and a notable eccentric. On July 6, radio signals were received in the United States informing the authorities that the 634-foot (193-m) airship planned to land on Roosevelt Field on Long Island, and that the tanks were low on fuel. Frantic efforts to have a landing crew ready at Montauk Point, the eastern tip of Long Island, proved unnecessary, as the ship appeared over Roosevelt Field.

a

b

Figure 2.2. The twin-engine Vickers Vimy, a British bomber of World War I, the first airplane to cross the Atlantic nonstop, on June 14, 1919. (a) Start in Newfoundland. (b) Crash-landing in Ireland.

A parachutist jumped from the airship. It was not General Maitland, who had been known to jump at the slightest provocation, sometimes preceded by a parachute with his personal effects, but rather a crew member who was to give instructions on how to handle the ship. Note that this trip was made from east to west against the prevailing winds, a feat that was not accomplished by an airplane until 1929 after Lindbergh's famous flight from New York to Paris. Three days later the R-34 began its return trip, covering 3,200 miles (5,100 km) in seventy-two hours and fifty-six minutes. This accomplishment resulted in the United States purchasing large dirigibles from Britain.*

Crossing the Atlantic—and later the Pacific Ocean—remained for years the goal of many aviators. Some failed before or at takeoff; others were fortunate to be rescued at sea; still others were lost. It was not after all an easy feat to perform: there is no route across the Atlantic Ocean that does not require flying more than 2,000 miles (3,200 km) over open water.

The most spectacular of the successful Atlantic crossings in the public mind was of course the solo flight of Charles A. Lindbergh on May 20 to 21, 1927, in the Ryan airplane *Spirit of St. Louis.* Lindbergh, a seasoned pilot of the U.S. Air Service who had flown the mail in every kind of weather, decided to make this flight in a single-engine monoplane.

He pitted himself against well-financed airmen such as Richard E. Byrd, who had extensive long-distance experience with a Fokker Trimotor, which required a crew of three. Lindbergh, who had carefully studied all previous attempts, determined that his unique solution would provide the safest way to success. His originality led to the view that he was a daredevil—remember that the Wright brothers were themselves misjudged as tinkerers—while in reality he was a calculating reasoner who based his solution on avoidance of the errors that he perceived in the ideas of his competitors. In particular he made careful calculations of weight and fuel load.

Poorly funded, he put his trust in the small Ryan Company of San Diego, California, proposing a larger version of their existing monoplane to be equipped with the special features needed for his trip. A gifted young Canadian engineer employed by the company, Donald Hall, set to work with him, and the entire group of workers at the Ryan Company devoted their efforts to the project.† In the amazingly short period of 60 days, the *Spirit of St. Louis,* shown in Figure 2.3, was completed and tested.

The aircraft became a "flying gas tank" to store sufficient fuel to cover the distance of 3,600 miles (5,800 km) roughly measured on a great circle on a globe at a local public library. Lindbergh chose the new Wright J-5 radial engine delivering 220 HP (164 kW). The wingspan of the earlier plane had to be lengthened, and other changes had

---

*See John Toland's book cited in App. 4.

†Much of the following is based on William Wagner, *Ryan the Aviator* (New York: McGraw-Hill, 1971).

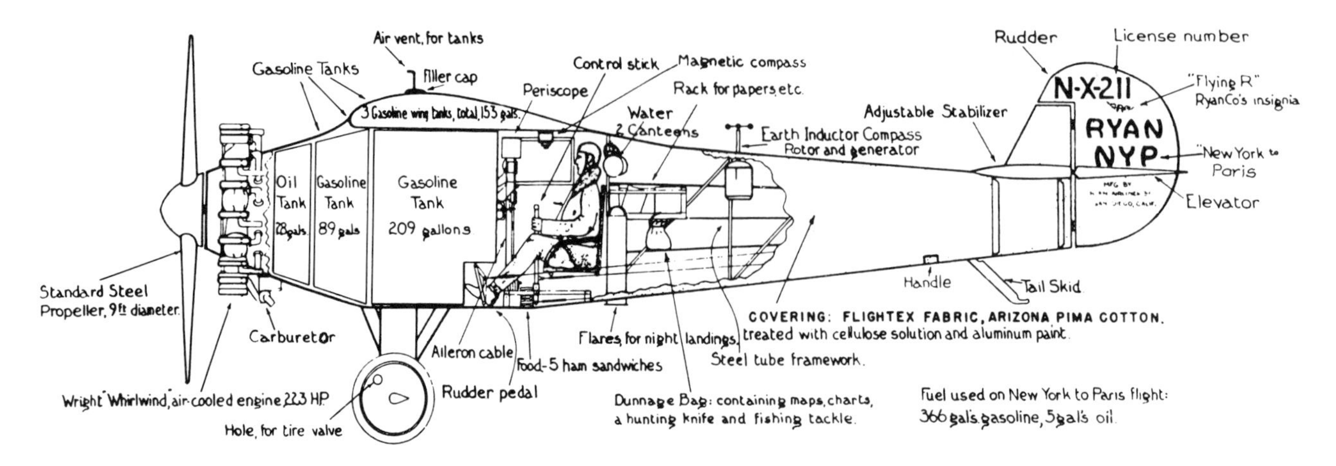

Figure 2.3. *A schematic side view of the* Spirit of St. Louis, *the Ryan monoplane in which Charles A. Lindbergh soloed from New York to Paris on May 20 to 21, 1927.*

to be made that inadvertently led to marginal stability of the aircraft (see Chapter 9.1), a situation similar to that of the Wright Flyer. Lindbergh commented that this fact would keep him awake! He closely watched the work on the aircraft and debated every smallest detail, flew its predecessor, and was considered a master pilot by designers and workers alike.

On May 8, 1927, two French pilots, the famous World War I ace Nungesser and Coli, took off from Paris. Lindbergh noted that this was the first time a plane loaded for the distance between Paris and New York had actually taken off. Both pilots disappeared in the Atlantic. Sensing the increased competition, Lindbergh took off from San Diego for New York on May 10. At New York, others as well as Lindbergh were ready to go on May 20, the day Lindbergh was set for his transatlantic flight. Byrd shook Lindbergh's hand and wished him well, and at 7:51 A.M. the *Spirit of St. Louis* started to roll. Some 451 gallons (1,700 liters) of fuel had been loaded, and the aircraft barely cleared a tractor sitting at the end of the runway.

The rest is history. Lindbergh crossed the coast of Ireland close to the point that he had aimed at by dead reckoning. At 10 P.M. on May 21, he circled Le Bourget Field in Paris and landed with a huge crowd assembled to greet him. He had flown thirty-three and a half hours at an average speed of 122 mph (196 km/h).* Something over 80 gal-

*We have used mph and km/h to give the speed of aircraft. However, this speed is often indicated in the books you may see by *knots,* a measure that has been taken over from the mariners. One knot is equivalent to the speed of one *nautical* mile per hour. The nautical mile is defined as the length of

lons (300 liters) of fuel were left in the tanks. Instantly, Lindbergh—an exceptionally appealing personality—became a hero to the world; he received the Congressional Medal of Honor and prize money ($25,000 had been offered eight years before), and he began an active career as a public spokesman for aviation. Toward the end of his life he devoted himself to environmental causes.

It took more than a year until the first east-west flight across the Atlantic succeeded where Nungesser had failed. The German Herman Köhl flew from Ireland to Labrador with a crew of two (von Hünefeld and the Irishman Fitzmaurice). They were followed in 1930 by two Frenchmen, Costes and Bellonte, who were the first to retrace Lindbergh's route in the opposite direction, flying from Paris to New York in slightly over thirty-seven hours in a Bréquet 19 aircraft. Also in 1930, a huge seaplane, the Dornier *Do X* pictured in Figure 2.4, flew in several stages from Germany to New York. This pioneering flight with a large crew and passengers foreshadowed commercial transatlantic air traffic. The *Do X* with its twelve engines had three decks, and at one time it carried 169 passengers on a short excursion flight. The first successful solo flight from east to west was carried out by a woman, Beryl Markham. In September 1936, she took off from England and, running short of fuel, crash-landed in Nova Scotia over twenty-one hours later.

In our selective narration, we now turn to the lesser-known early feats of polar aviation. Richard E. Byrd (1888–1957), who shook Lindbergh's hand before he took off, soon followed him in June 1927. Byrd flew the Atlantic in a Fokker Trimotor airplane with three crew members. After forty-two hours they crash-landed on the coast of Brittany. A pioneer aviator and polar explorer, Byrd had flown over Greenland in 1924, an experience that inspired him to try for the North Pole. In 1926, acting as navigator with Floyd Bennett as pilot, he started from King's Bay in Spitsbergen, the closest practical jumping-off point to the pole. The round trip to the top of the earth took fifteen and a half hours. Byrd also received the Congressional Medal of Honor for his pioneering effort. Only many years later, in 1970, was it determined that he probably had missed the pole.

By the late 1920s, Byrd could count on substantial financial support for his next expedition, a trip to the Antarctic continent. A base called Little America was established on the Ross ice shelf, from which Byrd explored previously unseen territory by air. In November 1929, again with Byrd as navigator and with Bernt Balchen as pilot, a successful round trip to the South Pole was made in fourteen hours. In

---

one minute of arc on a great circle of the earth. It is 1,853 m, versus the statute mile of 1,609 m. The knot therefore indicates a speed that is 1.15 times greater than mph.

a

b

Figure 2.4. The Dornier seaplane Do X *crossing the Atlantic from east to west in 1929. (a) Start in Germany. (b) Arrival in New York City.*

1930, Byrd became a rear admiral in the U.S. Navy's arctic service, and he led additional expeditions to this barren land. Byrd paved the way for flying under the extreme weather conditions in this region, where exploration is unthinkable today without aircraft, including helicopters.

The Norwegian Roald Amundsen (1872–1928) had been the first to reach the South Pole, on December 14, 1911. With a small group of good skiers and dog sleds, he made a daring dash and returned with all hands. Only thirty-four days later he was followed by the British Captain Robert F. Scott and four companions, who pulled their heavy sleds themselves and tragically perished on the return journey.

Amundsen understood early what aviation could do for polar exploration. In 1925, together with the American explorer Lincoln Ellsworth (1880–1951), he set out with two airplanes to fly to the North Pole. (Recall that Byrd's flight took place in 1926.) One aircraft malfunctioned, and both made an emergency landing on the ice, not far from the pole. The two Dornier Wal (whale) amphibious airplanes landed successfully on the rugged ice cover of the Arctic Ocean. One airplane, shown in the photograph of a diorama (a model scene) in Figure 2.5, was repaired. It then took several weeks of extreme effort to hack a runway in the pack ice, from which Amundsen and Ellsworth took off with four additional people in their overloaded airplane. They made it back to safety. (A year later the same type of aircraft crossed the South Atlantic.)

The following year, Amundsen and Ellsworth teamed up with the Italian airship captain Admiral Umberto Nobile, who piloted their dirigible *Norge*. They flew from a makeshift hangar on Spitsbergen—the traditional starting point of polar ventures—and crossed the en-

Figure 2.5. *A diorama of the Dornier Wal's emergency landing site on the polar ice during Roald Amundsen's arctic expedition in 1925.*

tire Arctic Sea to reach Nome, Alaska, via the North Pole. The same feat had been tried by the daring Swedish balloonist Salomon Andrée, who had led the first arctic flying expedition. Andrée and his comrades started in 1897 from the same location in their open basket, with a long rope trailing on the ground to give some directional control. A messenger pigeon returned to Sweden, but the expedition was lost until 1930, when its remains were found on a barren island northeast of Spitsbergen. The balloon had been forced down on the ice, and the group had tried unsuccessfully to march back.

The ambitious Nobile wanted to repeat the flight over the pole and pursue further arctic exploration. He mounted an Italian expedition with the dirigible *Italia* in 1928. The airship crashed north of Spitsbergen; a portion of the hull broke off and disappeared with some of the crew, while others, including Nobile, survived on the ice to be rescued finally. Although Amundsen, the experienced polar explorer, had not had a harmonious relationship with Nobile on the *Norge* flight, he took off immediately with his Dornier craft to search for the *Italia*. The search plane disappeared in the Arctic, and no trace of the mission has ever been found.

Much has been left out of our narrative, such as long-distance flights, coast-to-coast flights, and flights across the Pacific Ocean—in part with large seaplanes. I must, however, mention Amelia Earhart (1897–1937), a tragic figure among the great pilots. She crossed the Atlantic as a passenger of Wilmer Stultz and Louis Gordon in 1928, and in 1932 was the first woman and second person after Lindbergh to fly alone across that ocean. A solo flight from Honolulu to California followed in 1935. With a copilot, Frederick J. Noonan, she attempted to circle the earth, but her airplane was lost in the Pacific Ocean between New Guinea and Howland Island. The disappearance of this famous flier caught the imagination of many. It is still an unsolved mystery.

Transpacific mail service was introduced in 1934, and passenger service followed in 1937 with the *China Clipper.* The Boeing B-314, called the *Yankee Clipper,* provided the first scheduled Atlantic service in 1939, carrying seventy-four passengers. Either the northern route via Newfoundland or the southern route to the Azores was flown. Other aircraft were introduced, as described in the literature referred to in Appendix 4.

A multitude of competitions and air shows took place, where speed and altitude records were set and stunt flying was perfected. In fact, right after the First World War flying became a popular spectator sport. Barnstorming—a term taken from itinerant actors' troupes—here meant literally crashing into buildings in front of an appreciative audience. Pilots flew under bridges and raced around pylons, and the art of one-on-one dog fights of the war led to flying loops and other complicated figures. The daring flights led to advances in design.

Figure 2.6. The Hindenburg *disaster at Lakehurst, New Jersey, May 6, 1937.*

Again, discussion of the development of the many types of aircraft for civilian or military applications goes much beyond our scope. However, the characteristics of a few modern airliners, including the Douglas DC-3, the Boeing 747, and supersonic craft, will come up later in conjunction with aerodynamics. For more detailed information, aircraft buffs are referred to the widely available literature.*

Let us return once more to dirigibles. Huge airships provided regularly scheduled transatlantic flights between Europe and the Americas many years before airplanes had advanced to that point. This is a separate topic, well described by Harold G. Dick,† who himself made twenty-two transatlantic crossings on zeppelins.

The days of the large dirigibles—they had cabins to sleep in and pianos in their lounges—came to a dramatic end on May 6, 1937. The largest airship of them all, the German *Hindenburg,* arrived from Europe at Lakehurst, New Jersey. Preparations were being made for the landing; lines had been thrown out for the ground crew to guide

*See, for example, the books by Roger E. Bilstein and Walter J. Boyne cited in Appendix 4.

†See the books by Dick and Robinson and by Toland cited in Appendix 4.

Figure 2.7. Gossamer Condor, *flown by Bryan Allen and designed by Paul MacCready. Photograph taken during the first successful traverse of the Kremer course by a human-propelled aircraft, on August 23, 1977.*

the ship to its mooring mast. Something happened to cause an explosion—possibly an electrostatic discharge set off a spark. For reasons not fully understood, the hydrogen gas that provided the buoyancy for the huge craft (see Chapter 3) ignited. A photograph of the *Hindenberg* disaster, taken at the moment when the tail came crashing down, is shown in Figure 2.6. Remarkably, a relatively large number of crew and passengers survived the accident, but regular transatlantic airship service was never resumed. Service by airplanes slowly became a reality, and the giant airships that could have been built using the safe light gas helium were not constructed.

The final episode of our highlights of aviation, which took place in the 1970s, is a truly exceptional event, the first extensive flight of a human-powered aircraft. We saw in Chapter 1 that many such attempts had been made in history using the ornithopter concept, until at last Cayley early in the last century destroyed this notion. Converted sailplanes were tried occasionally in this century, but it remained for Paul MacCready to combine modern aerodynamics, extremely light materials, and a large dose of intuition and imagination to bring the old dream to life.

A specific incentive was provided by the offering of a prize, not the first, as we have seen, to prompt progress in aviation. The British industrialist Henry Kremer in 1959 established a prize for a human-powered airplane that could traverse a tortuous course involving an unassisted takeoff, 10-ft (3-m) hurdles, turns, and other challenges. In 1976, the prize money was raised to £50,000. By that time about fifteen human-powered aircraft had actually flown, but so far the Kremer course had proved to be too difficult for them.

MacCready saw the solution in a large, slow, lightweight (built with carbon-filament materials), wire-braced plane with a large propeller in the back, pedaled by a pilot weighing about 140 lbs, who must be a good cyclist.* After about four hundred test flights, an airplane weighing only 70 pounds with a 96-ft (29-m) wingspan emerged. (For more on human flight see Chapter 8.4.) On August 23, 1977, the *Gossamer Condor,* flown by Bryan Allen, successfully traversed the Kremer course set up at Shafter, California. It took roughly seven and a half minutes for the prize flight. A photograph taken on this epoch-making occasion is shown in Figure 2.7.

This remarkable event signaled progress in aviation at very low speed at a time when high speed was the rage, with commercial airliners flying in the stratosphere close to the speed of sound and the Concorde moving even faster at Mach 2 (see Chapter 10.2). Will we ever pedal through the air at twenty miles per hour or so on our way to work? We shall have to leave this tantalizing question unanswered. It is time to turn our attention to the science of flight, beginning with a consideration of the medium in which flight takes place.

---

*Paul MacCready, "The Gossamer Condor," *Astronautics and Aeronautics* 17(6):60 (1979).

Chapter Three

The Nature of Liquids and Gases

It is true that nature begins by reasoning and ends by experience; but, nevertheless, we must take the opposite route. We must begin with experiment and try through it to discover the reason.
LEONARDO DA VINCI (1452–1519)

3.1 Description and Properties

Before turning to flight itself, we need to discuss a number of fundamental ideas in order to lay down a scientific basis for aerodynamics. In this chapter we will learn about some of the physical properties of *fluids,* a term that includes both gases and liquids. This broad expression is used, as we will see later, because the motion of objects in air and in water obeys identical laws until their speed approaches the speed of sound. Occasionally this discussion—just like others to follow—may appear to be unrelated to the behavior of an airplane. Yet concepts such as pressure and phenomena such as the static behavior of water are relevant to the instrumentation used in wind tunnels and on airplanes.

Common sense tells us that the motion of fluids depends on their inherent makeup—honey and water behave differently—and in this chapter we will look at some features of fluids of interest to us. This will be followed by a discussion of the earth's atmosphere, the environment in which we fly. At that point the stage will be set for the field of *fluid mechanics,* the behavior of fluids at rest and in motion, on which aerodynamics is based.

Matter appears all around us in its three phases—solid, liquid, and gaseous. A solid substance offers resistance to a change of shape. To bend a steel rod you must exert a force; if you let go, it snaps back to its original position if you didn't bend it too far. If you bend it more, it remains in a new shape, or it may even break. In contrast, *liquids and gases do not offer resistance to a change of shape.* There are obvious differences between liquids and gases. For example, a gas such as air fills a container uniformly, and it cannot have a free surface like that of a lake. In addition, certain physical properties are obviously different. But the streamline patterns of gases and liquids—the lines of flow by which we visualize the motion of air or water about an automobile or a submarine, an airplane or a dolphin—are indistinguishable. Consequently, *both liquids and gases are called fluids.*

The properties of liquids and gases (color, taste, viscosity, optical behavior, etc.) depend on their diverse molecular structures.* The molecules of liquids are much more tightly packed than those of gases, and their motion is restricted. A gas molecule (or atom), on the other hand, moves freely in space until it bumps into another one that deflects its course. This substantial difference in packing and motion is reflected in the density, ρ (rho), or mass per unit volume of the two kinds of substances.

This can be seen in Table 3.1, which gives density and other properties required as we go along. These properties are shown for water and air, the fluids we primarily deal with. For comparison, other liquids of general interest are added. Mercury is still used for manometers, and the properties of castor oil are representative of typical lubricants. The *dynamic viscosity,* μ (mu), is a measure of the internal friction of fluids. This property will later play a major role in our understanding of fluid flow and aerodynamics. The *kinematic viscosity,* ν (nu), is defined as the ratio of dynamic viscosity to density, μ/ρ. For the flow of incompressible fluids, dynamic viscosity and density are constant. Consequently the kinematic viscosity is a constant, simplifying our description of such flows. It is important to note that if

---

*Before continuing, the reader ought to be familiar with Appendix 1 (elementary algebra) and the first two sections of Appendix 2. In the following, Greek letters will also be used, as is customary in this field. Wherever such a letter first appears, it will be followed by its spelled-out version—e.g., α (alpha). The full Greek alphabet can be found in Table A2.1. The reader who is not familiar with the metric system should also study Appendix 3. The reader should also understand the distinction between *dimension* and *unit.* For example, *length,* a dimension, can be expressed by the unit *foot* or *kilometer.*

TABLE 3.1. DENSITY, ρ (RHO), DYNAMIC VISCOSITY, μ (MU), KINEMATIC VISCOSITY, ν (NU), AND THEIR DIMENSIONLESS RATIOS WITH RESPECT TO AIR. IN THESE RATIOS THE SUBSCRIPT I DESIGNATES THE SUBSTANCES LISTED IN THE FIRST COLUMN. (VALUES APPLY TO P = 1 ATM = 14.7 PSIA; T = 20°C = 68°F.)

Substance	ρ kg/m²	μ (Ns)/m²	ν m²/s	ρ_i/ρ_{air}	μ_i/μ_{air}	ν_i/ν_{air}
water	998	1.00×10^{-3}	1.00×10^{-6}	825	55.0	6.67×10^{-2}
air	1.21	1.82×10^{-5}	1.50×10^{-5}	1	1	1
castor air	960	0.986	1.03×10^{-3}	793	54,200	68.7
mercury	13,500	1.57×10^{-3}	1.16×10^{-7}	11,200	86.3	7.73×10^{-3}

the word *viscosity* is encountered, it usually refers to the dynamic viscosity. These properties essential to our field will become much clearer once we apply them later.

Also shown are *dimensionless* ratios, which are obtained by dividing the value of a given property by the value of the same property for air. We find that water is about 800 times as dense as air, and liquid mercury—a metal—is 11,000 times as dense. The relatively tight packing of their molecules makes liquids practically *incompressible*—that is, an application of pressure does not change the volume of a liquid. Air and other gases can readily be compressed; you compress air when you use a bicycle pump.

Because molecules in large numbers in a system can be viewed as tiny bits of matter packed closely together, we deal with a *continuum*, a model of gases and liquids which assumes that fluids are indivisible. (Physicists and engineers frequently use the word *model* to designate a theoretical construct that reproduces the essential features of a complex physical situation. A model provides guidelines to problem solving; we shall soon encounter a model of our atmosphere.) The continuum of indivisible fluids governs our daily experience. When you see the wind move clouds or you drink a glass of water, you are regarding fluids as continuous substances. The molecular structure determines the particular properties of the fluid, but in fluid dynamics we need to concern ourselves only with bulk behavior.

The validity of this model becomes clearer when we take a look at the immense number of molecules involved in a fluid process. Sir James Jeans* provides a striking example by way of demonstrating that flows such as that about a wing should be treated by the laws of mechanics rather than the laws of molecular interactions. Jeans's example is as follows: There are about 3×10^{19} molecules of "air" (largely nitrogen and oxygen) per cubic centimeter in a room at normal pressure and temperature. In one breath we inhale about 0.4 liter (a liter

*J. H. Jeans, *An Introduction to the Kinetic Theory of Gases* (Cambridge: Cambridge University Press, 1948).

is roughly equal to one quart), or 400 cm^3, containing about 10^{22} molecules. (The number 10^{22}, a one followed by twenty-two zeros, is inconceivable to our imagination. However, we must understand the meaning and the handling of exponents, which are discussed in Appendix 1.) The entire atmosphere of the earth is made up of about 10^{44} molecules. Consequently, the ratio of the total number of molecules in the atmosphere to the number of molecules in one breath, $10^{44}/10^{22} = 10^{44-22} = 10^{22}$, is equal to the number of molecules in that breath. If we further assume that the atmosphere has been well mixed by turbulent motion so that all molecules have scattered in the course of the last two thousand years, it follows that every time we breathe we inhale at least one molecule of the dying breath that Julius Caesar exhaled as he was murdered in the Roman senate house in 44 B.C.! No readily foreseeable development in computer technology will permit us to follow the detailed history of individual molecules present in such enormous numbers in the simplest aerodynamic process. Their individual behavior is obliterated, and the mechanics of bulk motion prevails.

3.2 Behavior of Liquids at Rest

Fluid statics ("statics" comes from the Greek word meaning to make something stand) is the branch of fluid mechanics in which we deal with the distribution of pressure, density, and other parameters in the absence of motion—that is, in a state of rest or equilibrium. A swimming pool without bathers, an untouched cup of coffee, and the atmosphere on a clear, calm day follow the laws to be discussed here and in the next chapter. The concepts of fluid statics help us understand how balloons are supported by air and submarines are supported by water, just as the concepts of fluid dynamics describe the motion of such objects.

To discuss statics quantitatively, we first turn to the concept of *pressure*. The weatherman quotes the barometric pressure; your automobile tires carry air at a certain pressure. Its value may be given in pounds per square inch; its dimensions are consequently force per unit area (F/L^2). In a fluid at rest, the value of pressure at a given point is independent of the direction in which you measure it. A tiny pressure gauge will read the same pressure at a given location irrespective of the orientation of its pressure-sensitive surface. If there were no gravity acting on a fluid at rest—a condition realized only in the environment of a space laboratory—the pressure would be the same everywhere in the fluid. The ancient Greek scientist Archimedes (287?–212 B.C.) of Syracuse in Sicily was the first to explain that in the presence of gravity "each part of a fluid is always pressed by the whole weight of the column perpendicularly above it, unless this fluid

is enclosed someplace or is compressed by something else."*
Archimedes also developed the *principle of buoyancy,* now called after him, which states that an object immersed in a fluid experiences an upward thrust or lifting force exactly equal to the weight of the fluid it displaces. This law explains why steel ships float and light-gas or hot-air balloons rise. The latter have a lift force acting on them because hot air is lighter than cold air; at constant pressure, the density of air decreases with increasing temperature. Archimedes' principle can be used to determine the weight of irregularly shaped bodies: the volume of a body is equal to the difference of the weight of the body in air and submersed in water divided by the specific weight of water. (The latter term will be explained shortly.) It is said that Archimedes used this principle to determine if the crown of the king of Syracuse was made of gold. Once Archimedes knew the crown's volume, he could determine its monetary value by comparing the weight of this volume with that of an equal volume of gold. It turned out that the crown was not made of pure gold, with gold being denser than most base metals, and the offending goldsmith was supposedly executed.

The fundamental equation of hydrostatics ("hydro" refers to water), which quantifies the effects of gravity on liquids in equilibrium, can be derived by isolating a vertical cylinder of liquid in an extended space. This imaginary cylinder will not change its position if it remains in equilibrium. The forces acting on the cylinder exactly balance each other, and only weight and pressure in the vertical direction of gravity are important. The hydrostatic equation

$$p_1 - p_2 = wh,$$

which is valid for *incompressible* liquids only, tells us that the difference between the pressure forces acting on the opposing faces of our imaginary cylinder of height h is equal to the weight of the column of liquid, and therefore the cylinder remains at the same location. The pressure at the bottom, p_1, is larger than that at the top, p_2, since wh must be positive. Here we introduce the specific weight, w, a new term closely related to density and defined by $w = \rho g$, where g is the acceleration of gravity ($g = 9.81\ \mathrm{m/s^2} = 32.2\ \mathrm{ft/s^2}$). While density, as we have seen, measures the mass per unit volume M/L^3, specific weight indicates weight per unit volume F/L^3, a more practical unit in daily life. The hydrostatic equation tells us that pressure increases with depth, as predicted by Archimedes. The cross-sectional area of the cylindrical column does not appear in the equation because it is un-

*Hunter Rouse and Simon Ince, *History of Hydraulics* (New York: Dover Publications, 1963).

changed. It takes an enormous pressure, such as that found at the bottom of the ocean, to squeeze water sufficiently to produce a change in its density. This is in contrast to the small but crucial effect of temperature on water. The highest density of water is attained at about +4°C. At the freezing point of 0°C, when ice begins to form, water is less dense. Thus ice floats on the water in a pond, and fish can survive at the bottom.

The increase of pressure with depth depends on the density or specific weight of a liquid, as indicated by the w in the equation above, which tells us that in a pool of a given depth, h, a liquid with a higher specific weight exerts a higher pressure, p_1, at the bottom. The pressure distribution in a lake starts with atmospheric pressure at the surface and increases linearly—it obeys a straight-line function—with depth. Diving into the lake, we can feel the increase in pressure in our ears. Now we know that for water with $w = 9.81 \times 10^3$ N/m^3 = 62.4 lb/ft^3, every increase in depth by about 10 m (33 ft) adds another atmosphere or 14.7 psi to the pressure, with one atmosphere (designated by atm) equal to the standard atmospheric or barometric pressure. The latter value at sea level is counted from zero pressure and designated by psia (pounds per square inch, absolute). A deflated automobile tire is already at atmospheric pressure, and pressures above that value are designated by psig (pounds per square inch, gauge). In aerodynamics, however, we normally deal with pressure difference or with absolute pressure. (To get a clearer understanding of these concepts, see Tables A3.1 and A3.3.)

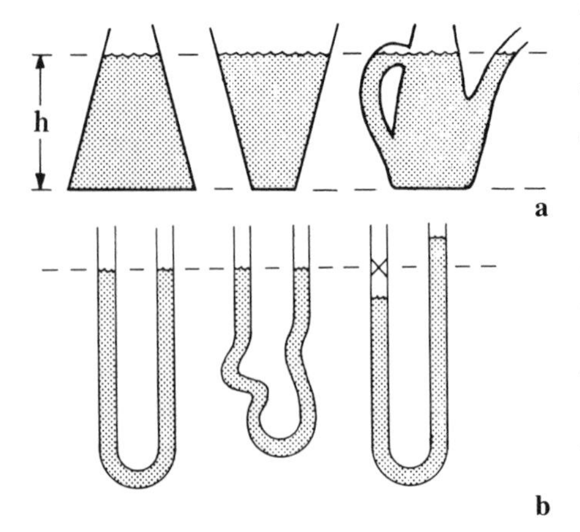

Figure 3.1. *(a) Pressure at the bottom of liquid-filled vessels is dependent only on depth. The shape of the vessel does not matter. The vessels shown have identical pressure at the bottom. (b) At left, two U-tubes exposed to the same pressure in both legs of the tubes. At right, a U-tube manometer with a valve in the left leg. The two legs are exposed to different pressures, and the difference between them can be read directly from the difference of the levels of liquid.*

The hydrostatic equation also shows that the increase in pressure for a given liquid is dependent solely on depth, and is somewhat surprisingly unrelated to the shape of the pool or other vessel. Thus the pressures at the bottom of the various containers shown in Figure 3.1a are identical. This observation is puzzling at first glance. It is called Pascal's *hydrostatic paradox* after the French mathematician and philosopher Blaise Pascal (1623–62). Say we have a large lake that has a dam at one end. The pressure at the base of the dam, where pipes lead to water turbines that drive electric generators, is given by our equation. The result depends only on the depth of the water at the dam; the surface area of the lake is immaterial in this context. Pascal also showed that the level of liquid in communicating (i.e., connected) tubes like the so-called U-tube (Figure 3.1b), whose open ends are exposed to the atmosphere, must be the same in both legs. If we apply pressure at the end of one leg of such a U-tube, the difference in pressure between that end and the open end can be measured from the difference in height between the two liquid levels. Many kinds of liquid-filled manometers or pressure gauges, including the mercury barometer, are based on this principle. For the barometer one leg of a U-tube is evacuated by a vacuum pump, and so the pressure in that leg is negligible. The difference in height of the mercury column in the two legs is now a direct measure of the atmospheric pressure; the weatherman announces that the barometer says 29 inches of mercury. It is essential to name the fluid with which the barometer is filled (normally indeed mercury), since its specific weight determines the height of the column for a given external pressure as shown by the hydrostatic equation.

The standard atmospheric pressure of 14.7 psia is substantial. Our bodies are constantly exposed to it, and it defines one of the many boundary conditions of man's evolution. The pressure exerted on surfaces often leads to enormous differential forces. For example, the atmosphere applies a force of over 2,000 pounds on a square foot of the exterior of a pumped-out vacuum vessel. In an aircraft the standard atmospheric pressure must be closely approached* for the comfort of the passengers. In turn, the aircraft must be sturdily built, because at high altitude at low external pressure, ruptures of doors or windows can be catastrophic. On the earth's surface, even small pressure differences owing to wind can slam doors or raise roofs, because the force acting on door or roof, which is given by the product of pressure difference and area, can be large.

Having reviewed the general properties of fluids, we are ready now to consider the earth's atmosphere in greater detail.

*Actually, cabin pressure in a modern transport aircraft operating at high cruising altitudes is normally somewhat less than 1 atm to ease the pressurization problem.

Chapter Four

The Atmosphere of the Earth

We live submerged at the bottom of an ocean of the element air, which by unquestioned experiments is known to have weight, and so much, indeed, that near the surface of the earth where it is most dense, it weighs (volume for volume) about the four-hundredth part of the weight of water . . . whereas . . . on the tops of high mountains it begins to be distinctly rare and of much less weight.

EVANGELISTA TORRICELLI (1608–47)

4.1 History and Composition

The atmosphere of the earth is a thin spherical shroud composed of a mixture of gases and retained by gravitational attraction. It extends to a great height, but conventional flight is possible only in its denser layers. In fact, about 99% of the total mass of the air is found below about 40 km (25 miles). Commercial airliners fly considerably below this height at roughly 30,000 to 50,000 ft (9 to 15 km), with the supersonic Concorde going to about 60,000 ft. General aviation* is concentrated at much lower altitudes. If you com-

*The term *general aviation* encompasses most private planes, short-haul feeder airlines, crop dusters, and the like, all moving at speeds far below those of the large commercial airliners.

pare these altitudes to horizontal distances on the earth's surface, the shallowness of tangible air becomes apparent; it is no more than the distance between a couple of rural towns. If the earth were a ball with a radius of one meter (roughly three feet), breathable air would extend only one millimeter (one twenty-fifth inch) outward. This thin layer of air makes life on earth possible. Planets such as Mars with a smaller mass and consequently a weaker gravitational attraction have a more tenuous atmosphere than ours, while yet smaller heavenly bodies such as our moon cannot retain a gaseous shell at all. This is in contrast to Jupiter, the largest planet in our solar system. With its mass 315 times that of the earth, its acceleration of gravity is about 3 times greater, and consequently heavy gases can be retained at high surface pressures. On earth, gases with the lightest molecules, like hydrogen, have flown off into space. By attaining escape speeds of over 11 km/s (7 mi/s), they can escape from the gravitational grip of the earth to fly into space, just like a spaceship on its way to the moon.

Earth's atmosphere shields us against excessive doses of ultraviolet radiation from the sun and against cosmic rays from outer space. Meteors burn up at the extreme temperatures generated during their high-speed atmospheric entry. With rare exceptions—the meteorites that actually hit the ground—their mass eventually settles safely as a fine dust. Potentially extreme differences in temperatures between night and day, like those encountered on the moon, are evened out by the atmosphere.

The atmosphere consists primarily of nitrogen (about 78% by volume) produced during the history of the earth by outgassing, a process through which rocks slowly give off stored gases. Nitrogen does not readily enter into chemical reactions because of the nature of its molecule N_2.* The life-giving oxygen in the air (about 21%) evolved in geological times since the formation of our planet about 5 billion years ago. The evolution occurred by a variety of processes whose details are still being debated. However, oxygen's current concentration spans a longer period than that of man's existence. The heavy nonreactive noble gas argon (A) appears in small concentrations. Carbon dioxide—produced in the biological cycle involving plant life—and water are found in varying amounts. Water exists as vapor—an invisible gas—or in liquid or solid form as rain or snow and ice. If all the water in the air precipitated, it would produce a layer only 2.5 cm (1 inch) deep on the earth's surface. Many trace

---

*N_2 stands for the diatomic molecule N-N. Chemical formulas such as N_2 (nitrogen), O_2 (oxygen), CO_2 (carbon dioxide), and H_2O (water) give a rudimentary picture of the chemical structure of the molecules.

constituents present in extremely small amounts complete the composition of air.

Again the aerodynamicist can safely treat air as the uniform, continuous fluid that we discussed in Chapter 3. Shown graphically in Figure 4.1, this gas mixture remains unchanged up to altitudes of about 100 km (62 miles), far exceeding the range of the highest-flying airplanes. Since the water-vapor content of the air varies substantially with the local weather, and its average value changes with the seasons and the location on earth, the standard atmosphere (see below) is taken as dry. Carbon dioxide varies to a lesser extent; it is, however, affected by local conditions, and an average value is therefore shown. The average value has been increasing slightly but steadily for the last 150 years or so. The increased burning of fossil fuels by power plants and in automobile engines, as well as other causes, may accelerate the CO_2 buildup. This may lead to the much-discussed greenhouse effect, since CO_2 traps the heat radiated by the earth. As we have already seen, air is much lighter than water (roughly 1:800), and the entire atmosphere has only 1/240 the mass of the oceans.

4.2 Structure

In contrast to water, air is a compressible medium (see Chapters 3 and 10). Consequently, its density changes with altitude in the gravi-

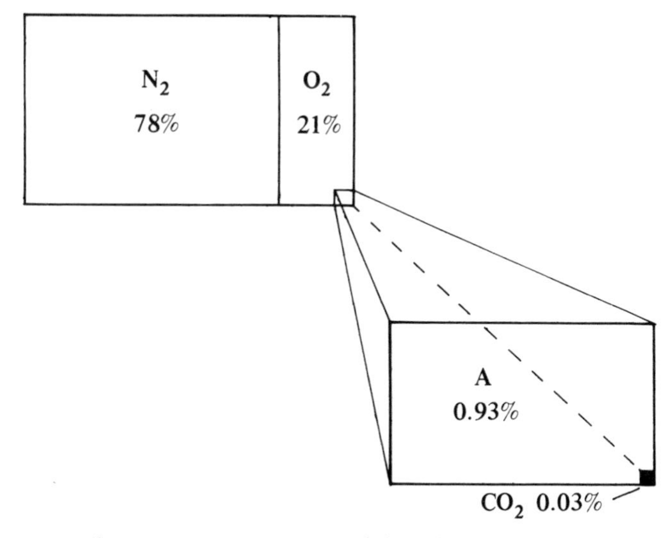

Figure 4.1. Schematic representation of the relative proportions (percent by volume) of the primary constituents of dry *air. Trace components such as neon, krypton, and helium are also present. An average value is given for carbon dioxide.*

tational field of the earth. The lower layers of air carry more weight than the upper atmosphere; in turn they are more compressed. Blaise Pascal,* whom we encountered before, and René Descartes (1596–1650)†—who both lived at the time of the first English and French settlements in North America—reasoned independently that the weight of the column of air creates the pressure to which we are subjected at sea level. They suggested that the pressure on a mountain ought to be lower than that in a valley. The same thought had been expressed by Torricelli and others, but it was Pascal who provided experimental proof of this fact. Experiments conducted in 1648 on a mountain near Clermont, France, in conjunction with barometric measurements in the valley fully confirmed the predicted decrease in pressure with height. Based on this observation in the seventeenth century, pressure gauges are still used to determine altitude. Such altimeters are found on every airplane, although they are now based on electromechanical devices.

We become physically aware of the change in pressure with height when we ride an elevator or climb a mountain. Only the hardiest mountaineers can climb Mount Everest, the earth's highest mountain (8,848 m = 29,029 ft), without carrying an oxygen supply, since the volume of oxygen in the air is linked to its pressure. Yet the summit of Mount Everest is substantially below the cruising altitudes of commercial airliners, which require pressurization to provide an environment sufficiently close to that at sea level for us to breathe properly (see Chapter 3).

The atmosphere has been much studied since the seventeenth century; daily and seasonal variations in certain properties, climatic and geographical variations, and effects of altitude have been noted. Instruments carried by mountain climbers have given new insights; weather stations have been permanently established at different locations and altitudes; ships have provided movable platforms for observations at sea; expeditions to all parts of the world, including the

---

*The SI unit of pressure, the pascal (Pa = N/m^2; see Appendix 3), is named in honor of Blaise Pascal, who contributed much to our understanding of fluid mechanics. Although used in science and engineering, this pressure unit has not yet become popular even among the practitioners of the field.

†The commonest system of coordinates that we use in this text is that of rectangular three-dimensional space; the coordinates are equivalent to the three lines that merge at the corner of a room. This system is called Cartesian after René Descartes, a graduate of a Jesuit college who later turned to the law. He made major contributions to science, mathematics, and philosophy. Such a varied career is characteristic of several early contributors to scientific knowledge.

polar regions, have contributed data on the weather; and instruments have been carried aloft by manned and unmanned balloons, airplanes, rockets, satellites, and space vehicles. Using the multitude of available observations, meteorologists have constructed a *standard atmosphere* possessing mean values of atmospheric composition, pressure, and temperature in a motionless and stable atmosphere which knows no seasons. Turbulent winds blowing near the surface, clouds and storms, jet streams at high altitudes, and global circulation patterns are all absent from these models of air at rest.

This composite picture of the properties of our air cover is a product of close international cooperation. The substantially different structure of the atmosphere between polar and tropical regions is ignored. There are slight variations in the listed values of the standard atmosphere as time passes and observations improve, but they need not concern us here since they are unimportant in the region of flight.

TABLE 4.1. PROPERTIES OF THE STANDARD ATMOSPHERE AS A FUNCTION OF ALTITUDE TO 30 KM.

Altitude km	ft	T(K)[a]	T(°C)	T(°F)	p/p_o	ρ/ρ_o
0	0	288	15	59	1	1
2	6,562	275	2	36	0.785	0.822
4	13,124	266	−7	19	0.615	0.666
6	19,686	249	−24	−11	0.466	0.539
8	26,248	236	−37	−35	0.352	0.429
10	32,810	223	−50	−58	0.262	0.338
12	39,372	217	−56	−67	0.192	0.255
14	45,934	217	−56	−67	0.139	0.186
16	52,496	217	−56	−67	0.102	0.136
18	59,058	217	−56	−67	0.0747	0.0996
20	65,620	217	−56	−67	0.0546	0.0727
22	72,182	218	−55	−67	0.0400	0.0531
24	78,744	220	−53	−63	0.0293	0.0384
26	85,306	222	−51	−60	0.0216	0.0280
28	91,868	224	−49	−56	0.0160	0.0204
30	98,430	226	−47	−53	0.0118	0.0147

Sea-level values are denoted by the subscript o.

$p_o = 1.013 \times 10^5$ Pa = 1 atm = 760 torr = 14.7 psia = 29.92 inches of mercury.

$\rho_o = 1.23$ kg/m³.

$g_o = 9.8$ m/s³ = 32 ft/s²; the value of the gravitational attraction at the surface of the earth remains practically constant in the range of this table.

H_o (scale height) = $p_o/(g_o\rho_o)$ = 8.4 km.

[a]Degrees Kelvin designate the scale of absolute temperature. T(K) is defined by T(K) = T (°C) + 273. You can see from the equation that absolute zero temperature occurs at −273°C.

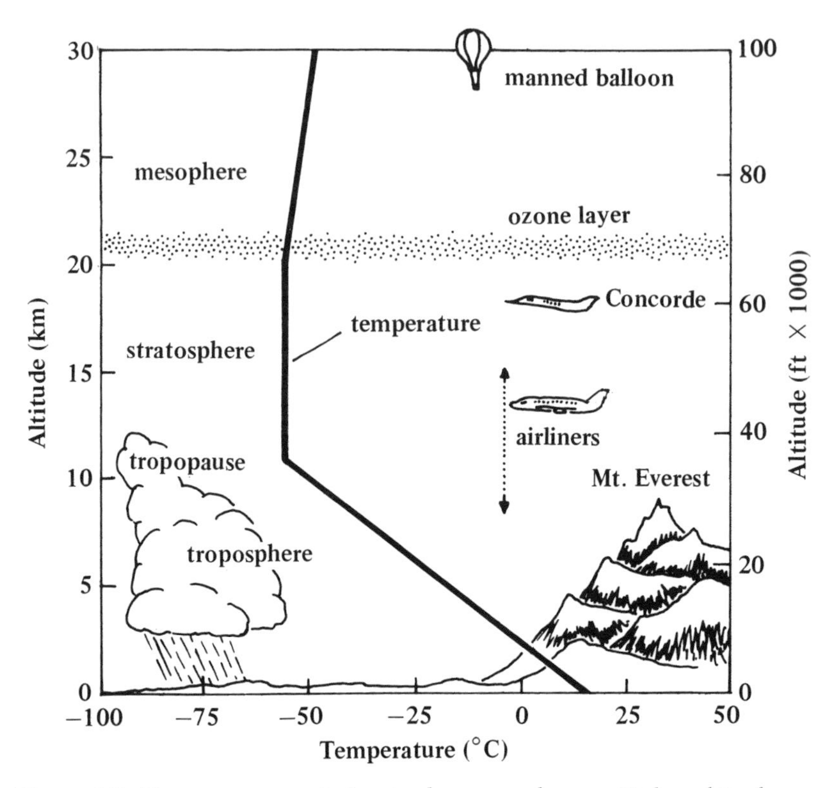

Figure 4.2. Temperature variation in the atmosphere to 30 km altitude and designations for different regions. Temperature values from Table 4.1.

Table 4.1 lists values of properties of the standard atmosphere. In addition to Celsius and Fahrenheit, temperatures are given in Kelvin, the absolute temperature scale, a measure that will later be required. Fig. 4.2 provides a graph of temperature versus altitude. The designations of various regions, the height of different phenomena, and the altitudes of practical flight are also indicated.

The lowest layer of the atmosphere, the one in which we live, is called the *troposphere*. The Greek word *tropos* means turning; turbulent air motion results in continual mixing, and the troposphere is host to much of what we call weather. The temperature drops linearly in the troposphere—that is, the decrease in temperature with altitude follows a straight line. The cooling of the air with increasing distance from sea level is about –6.5 °C per km (–3.6 °F/1,000 ft). This rate of temperature drop is called the *lapse rate*. The cooling effect—just like the drop in pressure—is readily noticeable if you ride in a cable car to a mountaintop. The simultaneous pressure drop follows a more complicated law that we will consider shortly. Pressure and temperature at altitudes equivalent to the highest mountain ranges are far too low for human life to be sustained. The troposphere extends to about 11 km, and at its upper edge—called the *tropopause*—there is a shift in the behavior of the temperature. Beyond the troposphere is the cold *stratosphere,* in which jet planes cruise. Here the temperature remains a constant –56 °C (–67 °F) for about 9 km; at higher altitudes the "thin," low-density air begins to become warmer because of chemical reactions that release energy, such as the production of ozone (O_3). At the lower altitudes of interest to us, however, no reactions occur.* This is one of the reasons that the composition of air remains unchanged in the region of flight. Taken together, the table and the graph reveal a complexity in the variations of temperature and pressure which is caused by the intricate interplay of these properties with gravity, radiation from the sun, and heating from the ground.

Looking again at Table 4.1, we note that the air pressure drops substantially with altitude. At 16 km (52,500 ft) the pressure is reduced to one-tenth of its value at sea level. Furthermore, we see in the table that the air *density* decreases with altitude. As noted before, in contrast to water, air is a *compressible fluid,* and the great weight of a column of air extending to the outer edge of the atmosphere compresses its lowest layer near the surface. It is remarkable that—as stated in the last chapter—the compressibility of air can be ignored in discussions of low-speed aerodynamics. But we are now forced in a discussion of the atmosphere to adopt a different view from the one we took in Chapter 3. The equations governing the static equilibrium of our extended atmosphere cannot be derived without more advanced mathematics, including the calculus, since the density appears as an additional variable. Pressure, temperature, *and* density must be considered simultaneously. (Later we shall find that fast aircraft pile up the air that is being compressed and form shock waves. The discipline of thermodynamics, treating the interplay of mechanical energy and heat, enters the realm of high-speed—not to mention supersonic—flight, as well as the structure of the atmosphere.) For these reasons the pressure as a function of altitude does not follow the simple law for liquids given in Chapter 3. The following may be skipped, and you can safely move on to Section 4.3. However, the behavior of atmospheric pressure is a beautiful example of the ever-recurring exponential function that ought at least to be mentioned.

A simple and practically important result appears if we assume the temperature to be constant in the atmosphere—that is, if we consider the atmosphere to be *isothermal.* A short while ago we referred to the drop in temperature with altitude—the lapse rate—and the assumption of a constant temperature appears to be a distortion of

---

*We will not consider smog, fires, explosions, volcanic eruptions, and the like that occur near the surface.

reality. Yet practically important results can be found, as we shall see. With this assumption we find that a fixed increase in altitude produces a fixed *ratio* of pressure drop. This fact is expressed by

$$\frac{p}{p_o} = e^{-h/H_o}.$$

In this equation $e = 2.72$ (see Appendix 1), and p_0 is the sea-level pressure. The symbol H_0 stands for the formula

$$H_o = p_o / (g_o \rho_o),$$

which has the dimension of length. A special name, the *scale height*, is attached to H_0. Inserting sea-level values in the last equation, we find for the scale height the value of 8.4 km (5.2 miles) as shown in Table 4.1. Setting $h = H_0$ in the pressure-altitude relation, we obtain $p = p_0(1/e) = 0.368\ p_0$. In other words, the atmospheric pressure drops to 37% of its sea-level value at an altitude of 8.4 km. In fact, for *any* altitude difference of 8.4 km the pressure drops by roughly 40%.

The isothermal and the real atmosphere differ remarkably little in pressure at low altitudes. Because of this fact, aircraft altimeters that measure only pressure can be calibrated directly in meters or feet from the equation above, which is often called the *barometric formula*. In the real atmosphere, however, distortions of the standard atmosphere appear. Pressure changes associated with changes in the weather cause periods of high or low pressure with respect to the average. They require a calibration of the altimeter according to the actual barometric pressure at sea level. A pilot receives this value during his meteorological briefing for his flight plan, and he can keep it current by radio.

Finally, a word on conditions at extreme altitudes that will have a bearing on our discussion later of the aerodynamics of space flight at the fringes of the atmosphere. Traces of atmospheric components characteristic of lower altitudes and other species produced by chemical reactions and ionization extend to great heights. Consequently, orbiting satellites, space shuttles, and other spacecraft leaving the atmosphere and returning to it encounter aerodynamic forces at much higher altitudes than those in which ordinary airplanes fly. The subfield of aerodynamics that deals with these phenomena is called low-density or *rarefied gasdynamics.** Even at an altitude of 1,600 km (1,000 miles), we still find atmospheric fragments, although their density is only about 10^{-15} of the sea-level value. But much closer to

*The word *gasdynamics* describes the fluid mechanics of compressible media (see Chapter 10).

the earth, where many satellites and the space shuttle circle it, interesting aerodynamic problems arise. Satellites often have elliptical orbits, in which the aerodynamic drag is uneven. Larger retarding forces act on a satellite at the *perigee* of its orbit, which is the closest point to the earth. Encountering resistance in the relatively denser air slows down the craft, and the result is a reduction of altitude at the *apogee,* the point of farthest distance from the earth. It is said that the orbit *decays,* and much effort is spent to predict this decay properly in order to anticipate the moment when the satellite will enter the dense atmosphere to burn up in a fiery trajectory.

At that final stage of its flight, the satellite becomes a man-made meteor, a luminous streak in the sky. It is occasionally possible for parts which do not vaporize to survive the aerodynamic heating (see Chapter 10) and hit the ground as meteorites. Satellites can be equipped with tiny booster rockets to restore the original apogee, yet this solution is only a short-term one. The clutter orbiting near and far in the space surrounding the earth will sooner or later cause fireballs and dust that will reach the earth's surface. However, the stationary communication satellites that beam television and telephone signals toward our houses are positioned at about 22,000 miles above us at the edge of outer space, where interplanetary conditions prevail. The pressure at this location is practically zero, outside of our interest in aerodynamics.

4.3 Global Circulation

We will finish this chapter with a brief note about the atmosphere in motion. The troposphere undergoes vertical air movement—for example, convection, an upward motion of air due to heating. This effect may alter the lapse rate and cause instability. The rising air gets colder. Once the moisture in the air reaches saturation at the dew-point temperature, it condenses on the huge number of *aerosols*—dust particles, salts, ions, and so forth—present in the air. The resulting movements of clouds, thunderstorms, and precipitation are part of the origin of local weather. The difficulty inherent in the meteorological prediction of local weather is all too apparent. Global circulation, however, concerns a more constant meteorological pattern, driven by the overall effects of the sun's radiation. Local weather, even the weather in a larger geographical area, results from a perturbation—or disturbances—superposed on the basic global pattern. Because of the higher position of the sun in the sky, more energy is delivered by radiation to the equatorial regions. In complicated interactions of pressure and radiation differences occurring at all latitudes, air rises near the equator and flows at high altitudes toward the poles.

Yet another force comes into play as well. The earth rotates, and motion on the earth is affected by its turning. A point on the equator moves at a speed of about 1,600 km/hr (1,000 mph), while a person standing on one of the poles rotates slowly about his axis once in twenty-four hours. The speed of motion on the earth due to its rotation is therefore dependent on latitude. The effects of the earth's turning can be visualized if you imagine a ball game played by people at various locations on a merry-go-round. The ball will trace trajectories that are surely not foreseen by the players. The first experiments and calculations of the mechanics of this behavior were carried out in 1835 by Gaspard de Coriolis in Paris. The force caused by the rotation of the earth, now called Coriolis force in his honor, pushes moving objects out of their straight path. Flows moving away from the equator turn to the east (to the right) in the Northern Hemisphere and westward in the Southern Hemisphere. The law applies to ocean currents as well: the Gulf Stream is the prime example found north of the equator.

Friction dominates air motion near the ground, where an atmospheric boundary layer exists (see Chapter 5). But at higher altitudes, winds that even out the imbalance of the circulation caused by uneven heating with latitude are the prime feature of interest to us. At the latitude of the North American continent we find the wind turning to the east after it arises in the south. The subsequent strong *westerly* winds* called *jet streams* may reach speeds up to about 400 km/hr (250 mi/hr). These powerful winds occur in the vicinity of the tropopause and affect the ground speed of airliners. (Here we refer to ground speed, the speed measured in relation to the earth's surface. The expression *cruising speed* refers to the speed of the aircraft with respect to the oncoming air. Ground speed equals cruising speed only in a perfectly calm atmosphere.) The powerful jet streams acting as head or tail winds affect flight times. They slow down east-west flights and speed up airplanes flying from California to the East Coast. The same effect appears over the Atlantic. Remember that the first transatlantic flights (see Chapter 2) were from west to east. This allowed the pilots to take advantage of prevailing tail winds. Flights from east to west, which would have to fight head winds, came later.

---

*Winds are designated by the direction *from* which they come.

Chapter Five

Air in Motion

Looking back on fifty years of aerodynamics research during the first half of this century, it appears most remarkable that the crude approximation which considers the air as an incompressible nonviscous fluid proved itself so valuable in solving many problems of practical aircraft design.

THEODORE VON KÁRMÁN (1881–1963)

5.1 Description of Movement: Kinematics

Aside from the first two chapters, we are following a sequence of topics that is quite traditional—except for the lack of mathematics—in the multitude of texts discussing fluid mechanics, the basis of aerodynamics. In this chapter we will close our discussion of the underlying science of flight. We have already seen that fluid mechanics teaches the behavior of fluids—liquids and gases—at rest and in motion. Only after mastering these fundamentals, albeit here without the use of higher mathematics, can we begin to understand aerodynamic drag and lift and what makes airplanes fly.

Observation of motion in liquids and gases—a river current, the wind blowing, the swirl of hot gases in flames, waves, clouds, rain—must have been among the earliest experiences of man. The harnessing of water for daily use and irrigation of crops (part of the transition from nomadic to settled life), advances in water transport from logs to rafts to streamlined vessels, and the making of spears and feathered arrows are engineering efforts that took place during prehistoric times. Egyptian, Greek, and Roman records tell us of the development of what is now called *hydraulics,* which is surely the oldest branch of fluid mechanics. The study of flow in rivers and open channels and motion in closed conduits such as pipes, and the design of water clocks, waterwheels, and windmills were at first largely empirical. A process of trial and error led to successful technologies. In contrast, modern aerodynamics involves the construction of models and the testing of them in wind tunnels, all based on scaling laws grounded in the so-called *equations of motion* of fluid mechanics. The solution of these equations contributes to the development of aerodynamic designs and tells us how to apply results found for small scale models to full-size airplanes.

The pioneers of fluid mechanics in the last century were building on progress in mathematics during the two preceding centuries, especially by the creation of a new branch of physics called *hydrodynamics.* The term *hydrodynamics* itself was introduced by Daniel Bernoulli (1700–82), a member of an extraordinarily gifted Swiss family of scientists and mathematicians. While still in his thirties, Bernoulli published his treatise *Hydrodynamica* (1738) in Latin, the scientific language of his time.

Fluid mechanics is now a mature field, indispensable to many areas of science and technology in addition to aerodynamics. It is involved in studies of the origin of the universe, the behavior of stars and galaxies (the latter look like the vortex of cream stirred in your coffee cup), the turbulent structure of the surface of the sun, and geophysical phenomena such as weather, ocean currents, and the earth's liquid metal core. Fluid mechanics helps explain the motion of fish, the flight of birds and insects, the dispersion of seeds in the wind, and the flow of blood in our body. Some dramatic recent technological advances, such as turbine-driven dental drills, chemical processes, and oil and gas pipelines, also involve fluid mechanics.

Before looking at the *equations of motion,* the mathematical descriptions of the physical behavior of fluid motion that are the basis of fluid mechanics, we must devise a descriptive framework in which to treat the dynamics of flows. There are two complementary approaches to studying flow around an object such as an airplane. If the airplane flies through the air—that is, it proceeds with respect to the atmosphere at rest—the flow moves relative to it. Similarly, if you ride in a car and stick your hand out the window, you feel the airflow moving relative to the car; aerodynamic forces such as drag and lift act on your hand. In the other approach, we mount a model of the aircraft on a support in a wind tunnel. We turn on a fan and blow a *uniform airstream* against the model. "Uniform" means that the air moves through a carefully designed inlet section to arrive at a constant speed in the *test section,* the location where the model is fixed to the wind tunnel. Researchers standing outside the wind tunnel view the model through a window. Both model and observers are at fixed positions in the laboratory; this is a most convenient way to study and describe air motion. The researchers measure surface pressures and determine by means of balances the forces of drag and lift on the fixed model. The two approaches described—free flight and the wind tunnel—provide identical testing environments. However, the great advantage of testing scale models in wind tunnels is that it eliminates danger to test pilots and the high cost of altering a full-scale airplane as research goes on. Furthermore, the wind-tunnel approach is the easier one for a discussion of aerodynamics, and in particular for drawing figures such as those in this text, in which the wind conventionally "blows" from left to right. The wind tunnel is not the only testing scheme used in aerodynamics: notes on other methods are found in Appendix 2.1.

Kinematics (from the Greek word meaning motion) is the branch of fluid dynamics that deals with the description of fluid motion, again including both liquids and gases. We must define a way to describe the spatial and temporal behavior of a fluid flowing by an object. In turn the motion of an object in a fluid at rest needs to be described. For example, the flight of a bird in space at different speeds must be put in mathematical form.

More simply, we ask how we can trace the movement of discrete pieces of fluid in space and time. Such pieces are often called particles or elements of fluid. (In the continuum description adopted here, it is important not to confuse such elements with molecules. The molecular structure of water and air, or any other fluid, is of no consequence once certain properties of the material are known, as discussed in Chapter 3.1.) If you throw bits of cork on the surface of a rapidly flowing creek, all of them will move along in the general direction of the current. A particular bit may follow a zigzag path and bob up and down with the motion of the elements of water around it, and at any given instant the value of its speed and its *direction* may change.

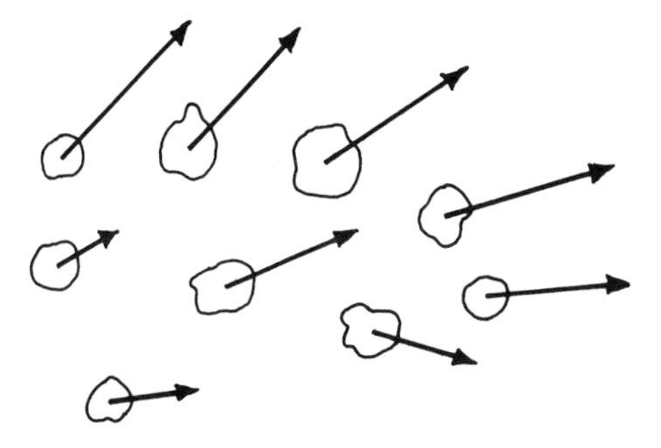

Figure 5.1. *Schematic representation of fluid elements in motion in an extended flow. The arrows represent vectors indicating flow direction and speed by their orientation and length.*

A schematic representation of this movement of pieces of fluid is shown in Figure 5.1. The *vector* arrows symbolize the motion of elements or particles of water at a given instant. Their direction indicates the instantaneous direction of the motion of the elements, and their length corresponds to the instantaneous value of velocity. In

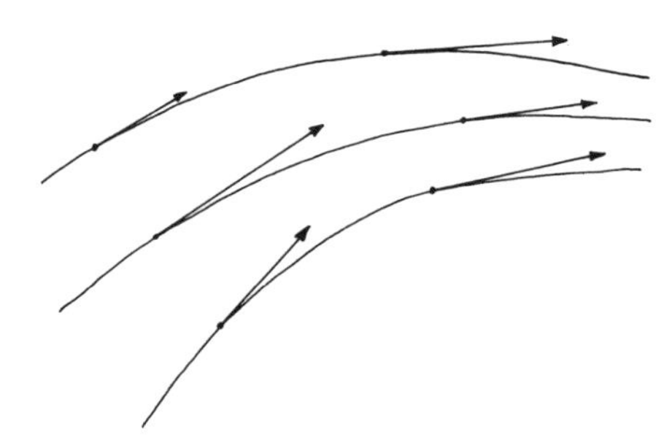

Figure 5.2. Streamlines are curves whose direction coincides at each point with the direction of the velocity of a fluid. The velocity is represented by arrows (vectors). In a steady flow, fluid elements move on streamlines.

principle, any element of fluid can move in any direction in space; such a flow is called *three-dimensional.**

We define a *streamline* as a curve that is at all times and at all locations tangent to the velocity vector, as shown in Figure 5.2. In simpler terms, a streamline has at every point in the flow the same direction as the local velocity. (This concept is not an easy one to visualize until we bring the time-dependent behavior of flow into the picture.) Streamlines are made visible in wind tunnels by introducing filaments of smoke. An example of actual streamlines about an airfoil or wing section is shown in Figure 5.3, where a wing is extended from wall to wall in the wind-tunnel test section at a right angle to the plane of the photograph. (The term *Reynolds number* given in the caption for the figure will be explained later.) Flows like this, called *two-dimensional* or *plane* flows, are much easier to understand than three-dimensional flows. The streamline picture will look identical wherever you cross-section the wing or, for another example, a weir over which water is flowing.

We complete the views of flow in space by defining the simplest situation, *one-dimensional motion.* For example, the flow in a pipe follows a single coordinate, the length of the pipe. This convenient description can be extended to motion in ducts of different cross section, such as a wind tunnel. Easily visualized, one-dimensional flow has many important applications in which an average, uniform speed at a given cross section can be defined.

In addition to the geometry or shape of the *flow field,* defined as a spatial view of the motion in an extended space, there is a temporal or time-dependent aspect of motion. So far we have implicitly assumed that the flow is *unsteady.* This means that we permitted continual changes of the flow with time. Such a situation is seen in the movement of clouds during a storm. We have considered a flow such as that in Figure 5.1 as a brief glimpse of the motion, a snapshot of the streamlines that at a later moment would look different. Unsteady flow occurs when a driver accelerates in order to attain a certain highway speed. An airplane taking off and a baseball being pitched also involve unsteady flow. A series of moving-picture frames of these actions would reveal rapidly changing streamlines. But let us now assume that on the average—little variations of flow speed about a mean value can here be neglected—there is no change of flow speed with time at a given point anywhere in the flow field. Such flows are called *steady.* If the flow is steady, velocity vectors such as in Figures 5.1 and

*Recall the Cartesian rectilinear coordinates (i.e., at right angles to each other) from Chapter 4.2; the particles in the flow may move anywhere in this space, like flies buzzing around in a room.

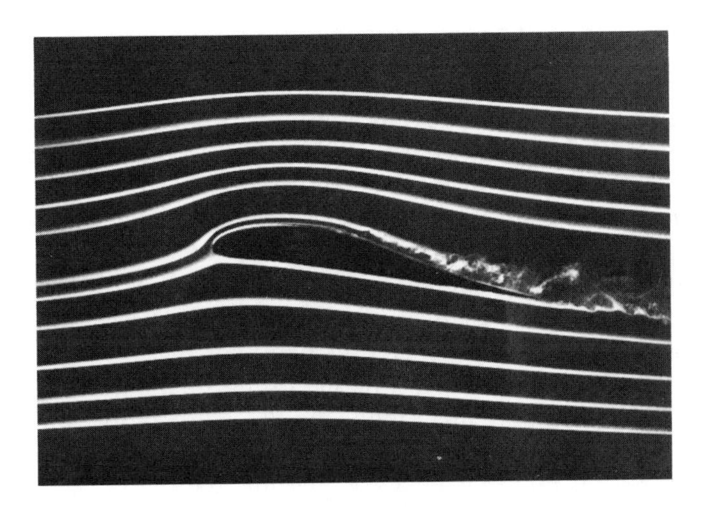

Figure 5.3. A photograph of the flow about an airfoil model suspended in a wind tunnel at an angle of attack of 10° and at a Reynolds number of 150,000. Flow is from left to right, and the streamlines are made visible by smoke filaments.

5.2 do *not* change direction or length with time; succeeding frames of a movie would all look alike. Obviously there will be different speeds at different locations, since the constancy of flow velocity refers only to fixed points. The flow shown in Figure 5.3 is in fact steady. But typically, the air moves faster at the top of an airfoil than below, and it comes to a full stop at the nose or leading edge. Yet the streamlines look alike at whatever time you photograph them.

It is indeed a happy circumstance that steady flows with their simpler behavior occur frequently, typified by an aircraft at constant cruising speed, a car at fixed highway speed, a ship on the high seas, or in fact the water flow from a faucet once it is turned on. Here we will deal largely with aerodynamics in the steady state, which will permit us to take advantage of the simple steady-flow description. Returning to the bits of cork on the creek and the filaments of smoke in the wind tunnel, we fasten our eyes on a given spot. Every new bit of cork or wisp of smoke at that spot moves like the previous one. Its journey traces that of its predecessors that have passed that spot in the flow. It may move more slowly or rapidly at other spots, changing direction in space, but each piece at the spot we are watching will show the same speed and direction. Recalling the definition of streamlines, we therefore note that in *steady flow, particle paths and streamlines are identical*. This fact takes a little contemplation for a better understanding: observe the clouds in a steady wind, the drift of leaves in the fall, or a waterfall, and try to fathom the kinematics of these flows in light of what we have said.

5.2 Conservation of Mass and Energy

Conservation laws pervade physics. Certain quantities are immutable: they may change their appearance, but their total sum remains constant. In fluid dynamics we deal with the conservation of mass and mechanical energy. In thermodynamics the total energy includes heat in addition to mechanical energy. When we come to high-speed flows and supersonic flight, we must add this aspect. Heat and mechanical energy are interchangeable (both have the dimension of work, force times length). The interchange of these two forms of energy has certain constraints that will not concern us here. For the aerodynamics of automobiles, ships, aircraft at low speeds, birds, and golf balls, thermodynamics does not enter. Following the conservation laws, mass cannot increase or disappear in a flow; it may alter its distribution, but its fixed total can always be accounted for, and the same is true of mechanical energy.

First let us consider the conservation—or continuity—of mass. For the remainder of this chapter we shall assume an *incompressible flow*, in which the density of the fluid remains constant. The assumption is obviously true for water (see Chapter 3); air too can be regarded as incompressible as long as flow speeds remain reasonably low.* This assumption is roughly valid as long as airplanes fly slower than about 350 km/hr (220 mph), or about one-third of the speed of sound. Since commercial airliners exceed 350 km/hr when cruising in the stratosphere, where the speed of sound is about 1,060 km/hr (660 mph), we will later have to deal with the effects of compressibility of air. But it is in fact remarkable that the aerodynamics of most small airplanes and the drag of golf balls, the flight of an arrow, and the flow about a submarine or a fish are based on identical laws.

At constant density, the conservation of mass is equivalent to that of volume. Take a *steady* flow, for which the flow speed at a given location remains constant, and construct a *stream tube* (Figure 5.4) by singling out a number of streamlines passing through the periphery of some arbitrarily chosen cross-sectional area A_1. Let us follow these streamlines downstream and designate a second cross section, A_2. Since all streamlines point in the direction of the flow, *mass cannot cross streamlines.* Consequently our stream tube can be viewed as a pipe with solid walls whose cross section varies in size. Whatever fluid volume per unit time enters area A_1 must come out at the same rate at area A_2. The steady flow of incompressible fluids through the stream tube is therefore fixed by the discharge rate, Q, given in terms of volume per unit time (L^3/T). Defining an average flow speed for each cross section in the one-dimensional flow picture, and calling it

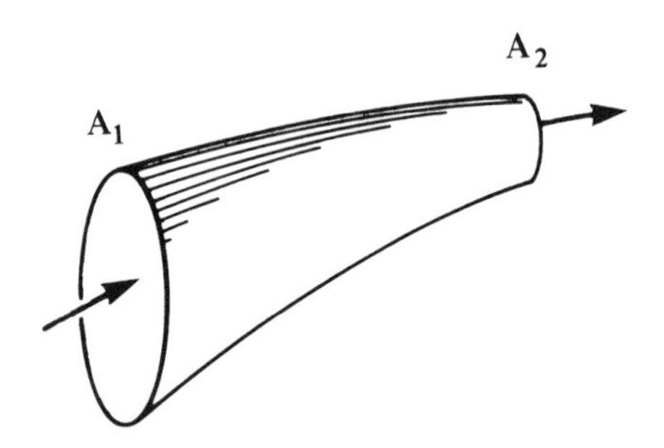

Figure 5.4. A variable-area stream tube made up of streamlines connecting areas A_1 and A_2 singled out in an extended flow that is not shown.

*Of course, no fluid or solid material is absolutely incompressible. Even a piece of steel changes its volume if we squeeze it hard enough.

u, the discharge rate is given by

$$Q = u_1 A_1 = u_2 A_2 = uA,$$

where the subscripts 1 and 2 denote the beginning and the end of the stream tube shown in Figure 5.4. In steady flow the discharge rate—or volume of flow passing through the tube in a given time—must, of course, remain the same at any cross-sectional area in the tube. It is computed from the product of flow speed and area; with the dimensions of $A(L^2)$ and $u(L/T)$, this product yields $Q(L^3/T)$. We multiply the volumetric discharge rate Q (often given in the units CFM or cubic feet per minute) by the constant density (mass per unit volume) to confirm that the mass passing each cross section of the stream tube per unit time is also constant. This fact expresses the conservation of mass as applied to a moving fluid. Unless we add or remove fluid in a given flow, the mass must remain constant.

But let us return to the last equation. Intuitively we believe flow to speed up in a constriction. In the stream tube in Figure 5.4, the streamlines converge toward a section A_2, and we expect the average speed of the steady flow to increase toward the smaller area A_2 in order to maintain the fixed discharge rate. From the last equation we find that

$$u_1/u_2 = A_2/A_1,$$

which demonstrates that, at a constant discharge rate Q, the average steady-flow velocity at a given location in the tube is inversely proportional to the area, as shown by the dashed line for flow speed in the converging-diverging duct in Figure 5.5. The increase of flow speed

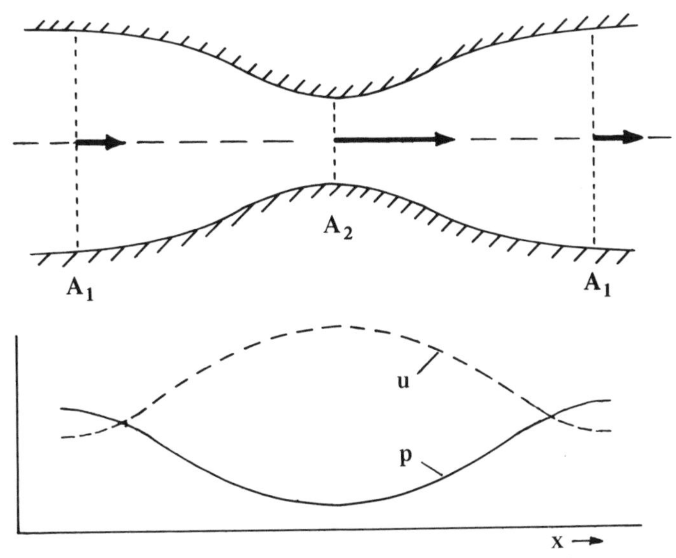

Figure 5.5. A converging-diverging passage in steady, one-dimensional flow. The arrows represent the average flow speed through the duct. Pressure and flow speed for ideal flow are shown below.

is directly related to the decrease of the cross-sectional area. This inverse proportionality is shown by comparing the velocity curve with the duct area in Figure 5.5. Since our duct is symmetrical about the middle at the smallest area, the flow speeds at the entrance and exit are equal. The flow speed is highest at the narrow section in the middle. Remember again that we are dealing with steady flow; the flow speed changes throughout the duct, but it remains constant with time at any fixed location. It is interesting to note that these simple facts apparently were not understood by the Romans. Still, their engineering expertise enabled them to build an elaborate system of aqueducts, providing more water for the city of Rome than is used even today.

The inverse relation of flow velocity and area is well known to us from experience. If we move slowly in an extended crowd of people toward a turnstile, our initial low speed picks up as the way narrows. Or take the traffic pattern on a highway. If all lanes but one are blocked ahead of us, traffic backs up in several lanes moving at low speed, while each car passes the obstruction rapidly.

After applying the law of conservation of mass to fluid dynamics, we must next consider the conservation of energy. As for mass before, if we single out a specific flow and do not add or remove energy (or work: see Table A2.2), the energy must remain constant. The dynamics of motion can be understood by applying Newton's law of force (see Appendix 2) to a moving, incompressible fluid, anticipating that the sum of the different forms of mechanical energy involved must be conserved. These conservation principles of energy are most easily described for the model of an *ideal fluid*. In such a fluid, the internal friction caused by viscosity, which acts to retard a change of shape of the liquid or gas, is neglected. We have an intuitive feeling for the property of viscosity (see Table 3.1)—for example, toothpaste is obviously more viscous than oil, oil is more viscous than water, and so on. Ignoring viscosity in the science of fluid dynamics appears to be a drastic assumption; however, in some common situations involving water, air, and other fluids of practical importance, friction plays a minor part. In automotive aerodynamics, for example, the surface friction can be ignored. Moreover, water and air have small viscosity values—in contrast to heavy oils, honey, and the like, for which viscous forces can never be neglected—resulting in small internal resistance to deformation by external forces (see Table 3.1). The usefulness of these ideas was succinctly summarized by von Kármán in the quotation that appears at the head of this chapter. Nonetheless, we shall quantify viscosity and describe its role in fluid dynamics later.

Neglecting friction, we consider first the *potential energy* stored in a mass of fluid owing to its position, z, above an arbitrarily chosen baseline z_0 in the gravity field of the earth. Water stored in a water tower has a higher potential energy than water on the ground, and it is potential energy that makes water run downhill. Next we have the energy represented by the force exerted on the fluid by its internal pressure, p. If you increase the pressure in a tube of toothpaste by squeezing it, you make the paste come out faster. Finally, the *kinetic energy* of the fluid is related to the square of the velocity of a fluid element. The fluid acts like a golf ball: if it moves faster it goes farther, because it has a higher kinetic energy. In the stipulated absence of friction, *the sum of these three energy terms must remain constant.* For steady flow of an ideal, incompressibe (ρ = constant) fluid, a derivation of the energy law yields

$$p_1/\rho = gz_1 = (1/2)u_1^2 = p_2/\rho = gz_2 = (1/2)u_2^2 = \text{const.},$$

where g is the constant acceleration of gravity (Table 4.1). This famous expression is called *Bernoulli's equation* (sometimes *theorem* or *principle*). It can well be regarded as the single most important statement on flow in incompressible media, because it explains a wide range of important phenomena.

Three energy terms counted from left to right—pressure, potential energy, and kinetic energy—are again written for two locations along a stream tube. They all have the dimension of energy per unit mass of moving fluid. The demand for dimensional homogeneity (see Appendix 2) is therefore met, as can be readily seen, for example, if we consider the potential energy, or gravity, term, gz. The dimension of the acceleration of gravity (L/T^2) times the height z above some baseline z_0 of the dimension L yields $(L/T)^2$. Recalling that energy and work have the dimension $M(L/T)^2$, we find the energy per unit mass, arrived at by division by M, to be $(L/T)^2$. This dimension is also that of the *square of velocity,* as is immediately apparent from the kinetic energy, the last of the three terms. You may wish to recheck all three energy terms in Bernoulli's equation in the same manner with the aid of Table A2.2.

To repeat, the total energy in a flow is calculated by summing up the three different forms of energy. It is this sum that remains constant in the flow, while the mode of energy may shift among the three different terms. For example, when a ball runs downhill or water rushes downward, losing altitude and picking up speed, potential energy is converted to kinetic energy. Other combinations and trade-offs are possible, provided the total of the energies remains constant. We can now link the laws of conservation of mass and energy in a variable-area stream tube by combining the equation for the continuity of volumetric discharge with Bernoulli's theorem.

We now have in hand the basic concepts of flows with small or no effects of internal friction; they relate area, velocity, and pressure to each other, accounting for gravity if necessary. *Decreasing cross-sectional area leads to higher speed, and higher speed causes lower pressure.* Let us return to Figure 5.5. Here the solid line in the graph at the bottom of the figure gives the pressure as a function of distance along the duct. The duct is horizontal, and with a constant value of z, the gravity term gz in Bernoulli's equation remains constant. With constant density, we can simply trade speed against pressure as we go along. The highest speed at the narrow section of the duct leads to the lowest pressure in the flow. Because of the symmetry of the duct we find equal pressures at the entry and exit, just as we found equal speeds. The flow pattern in the converging-diverging duct given by flow speed and pressure is now complete. It is based on the conservation of mass (or volume) and energy for a steady flow that is incompressible and inviscid (no viscosity)—that is, an ideal fluid. This last sentence reminds us that we have now acquired a substantial vocabulary in our field.

The tendencies shown in Figure 5.5 reappear qualitatively in many other interesting situations. Say you split the duct lengthwise along its center line. The lower duct wall can now be taken as the upper surface of the wing of an airplane. The distribution of pressure and velocity on the wing surface would look quite similar to that shown in Figure 5.5.

Besides being important in the study of aerodynamics, these relations of area, velocity, and pressure provide the key to understanding many devices utilizing the Venturi tube, a pipe with a constriction at which the lower pressure activates suction. The Venturi tube is used in carburetors, instruments that measure discharge rates in plants that handle liquids, siphons, waterjet pumps, paint sprayers, dental apparatus, and propulsion systems for boats. The wide range of applications of Bernoulli's equation extends to flows that are not confined to ducts and pipes. The flow over dams or weirs—including the geometrical shape of the flow, the behavior of flows in front of and around obstacles, and flows through sluice gates and cutouts—is an example we can now treat according to the principles developed here. We must always keep in mind the two facts—reduction in area speeds up a flow, and higher speed results in lower pressure. Closer to our subject, aerodynamics in a wider sense does not concern itself only with the outer flow about bodies. Aerodynamics extends to flows confined by enclosures. Typically, an aircraft is propelled by jet turbines, cooling systems are ducted, and liquids are metered. To all of these Bernoulli's equation applies as a useful approximation. The flight of a bird, the trajectory of a shooting star, and the lift of an airplane are all governed by variants of the same laws.

As a topic in fluid statics, Archimedes' principle deals with the way in which hot-air balloons rise because they are lighter than air. From these beginnings hydrogen- or helium-filled airships evolved, providing the first large-scale and long-range air transportation (Chapter 2). But birds and powered aircraft, both heavier than air, sustain themselves by producing aerodynamic lift, an upward force that counteracts their weight (Figure 1.1), and so Bernoulli's equation comes into play. Early technology made use of the reverse effect: forces exerted by wind on the blades of a windmill were made to turn the shaft. Windmills designed to produce power, once as commonplace as waterwheels, can still be seen in many parts of the world (Figure 5.6). In fact, aerodynamically improved wind motors are beginning to provide an important source of energy. In flight the propeller, a reverse windmill, pulls an aircraft forward. In Figure 5.3 we see that the streamlines about an airfoil section squeeze together to pass around the upper wing surface. Anticipating a discussion of the aerodynamics of lift, we apply our two basic rules to this flow. Because of the constriction of the stream tube, the airspeed is higher above the wing than ahead of it or below it. Higher speed implies lower pressure, and we expect the airfoil to rise.

Figure 5.6. *The Pakenham windmill near Bury St. Edmunds, England. This old grain mill has a rudder-like appendage moving on a circular track to point the vanes into the wind.*

In dealing with our primary interest, air motion, the effects of gravity are fortunately sufficiently small to be neglected.* This is a consequence of the substantial difference in specific weight (about 1,000:1) between liquids and gases. We see this difference in gravity effects when we observe a water jet from a garden hose bending toward the ground and an air jet that remains straight in every orientation. Taking advantage of this situation, we multiply each term in Bernoulli's equation by the constant density, ρ, and drop the gravity term to find for two locations 1 and 2 in the flow

$$p_1 = (\rho/2)u_1^2 = p_2 = (\rho/2)u_2^2 = p_o,$$

or in general

$$p = (\rho/2)u_2 = p_o.$$

We are now dealing with an even simpler version of Bernoulli's theorem. Here we have designated the constant of the previous expression p_o. This symbol stands for *stagnation pressure,* while the pressure, *p,* in the flow is called the *static pressure.* Return to Figure 5.3, and assume that small holes have been drilled at various places on the surface of the airfoil. Pressure gauges attached to them will measure the static pressure. We can also invent an imaginary pressure gauge that moves with the wind in a wind tunnel. This gauge will read the static pressure. Dimensional homogeneity (see Appendix 2.2) demands that the term $(\rho/2)u^2$ also have the *dimension of pressure,* a fact that you can verify yourself. This new term, which is called the *dynamic pressure,* is directly proportional to the square of the velocity—the kinetic energy per unit mass—since the density is constant. The stagnation pressure, p_o, represents the constant total energy in the flow. While we can trade between speed (via the dynamic pressure) and the static pressure, the sum of the two must remain constant and equal to the stagnation pressure. Again this statement is a consequence of the conservation of energy in this specific situation. The total energy remains unchanged.

What about the practical aspects and the measurement of p_o? If we drill a hole in the leading edge of the airfoil in Figure 5.3 facing the wind head on and again attach a pressure gauge, we will read p_o. The motion of the center streamline comes to a full stop at this point. With $u = 0$, we find $p = p_o$ from Bernoulli's equation. In practice the stagnation pressure is measured with a *Pitot tube,* an open tube oriented into the wind that you can see on the nose, wings, tail fin, or fuselage of any airplane. The Pitot tube is named after Henri de Pitot (1695–1771), who devised the first instrument of this type to deter-

*For liquid flows, gravity effects are absent in horizontal systems such as a pipeline on level ground; here z = constant.

mine the flow speed of a river. He dipped his tube into the water, and by estimating the water depth, he could derive an approximate flow speed. Pitot also measured the speed of a boat and anticipated a direct measurement of airspeed. Since static pressure is measured on the surface of an object, the determination of stagnation and static pressures may be combined in a single device known as a Pitot-static tube or probe, which is shown schematically in Figure 5.7. Two concentric tubes are joined at the tip, with the inner tube open to the wind. The nose is rounded to ensure smooth flow about the device. A number of small holes are drilled on a circumference of the outer tube. These holes sense the static pressure of the outer flow. The inner tube is exposed to the stagnation pressure, since the air comes to a full stop at its entry. The two separate tubes are connected to a differential manometer. Such a pressure gauge responds to the *difference* in pressure, just like the mercury-filled U-tube in Figure 3.1b. (On modern airplanes the pressure difference is determined by a variety of electromechanical devices that are often based on the deflection of a membrane between two chambers.) A well-designed Pitot-static probe is remarkably insensitive to the angle of the tube in relation to the wind, provided the angle remains relatively small. This is a useful feature on an airplane that may fly at an angle of attack, since it is difficult to point the device continually into the wind. Pitot-static tubes can also be used on ships and sailboats, where they are attached to the hull.

Now we know the difference between the stagnation and static pressures. The result gives the dynamic pressure, from which we can derive the velocity of flight. From Bernoulli's equation we have

$$u = [(2/\rho)\,(p_o - p)]^{1/2},$$

noting that the air speed is proportional to the square root of the difference between stagnation and static pressures, since the density is a constant. (The exponent 1/2 is used to indicate the square root of the expression on the right-hand side; see Appendix 1.)

The pressure difference measured by the differential manometer can be registered on a scale seen by the pilot that is calibrated in terms of flight speed. Using a Pitot-static probe with a differential manometer along with an altimeter (see Chapter 4), the pilot can now determine speed and altitude. Note how readily we found simple laws that permit us to find these two values essential to navigation.

A second determination of the stagnation pressure is important in wind-tunnel testing. As we have seen, an airplane model is suspended in the test section, where the required wind speed is attained by a compressor or fan. The test section is preceded by a large settling chamber, also called a plenum or reservoir, in which the flow speed is nearly zero. Therefore the static pressure in the reservoir is equal to p_o, now called reservoir or supply pressure. The latter term describes

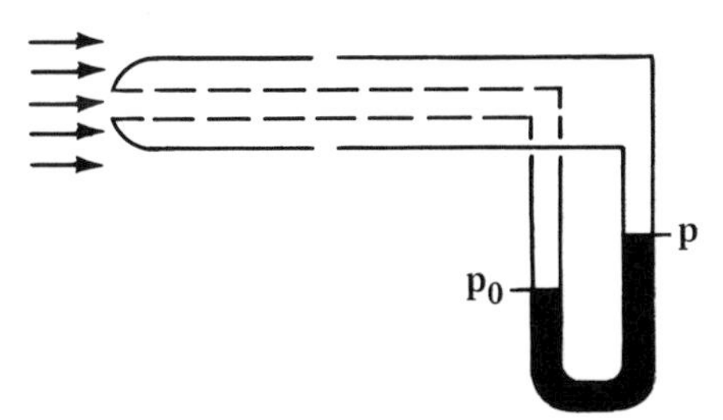

Figure 5.7. A Pitot-static tube with an attached differential manometer whose scale can be calibrated directly in values of air speed.

the condition at the wind-tunnel inlet. The sum of static and dynamic pressure in the flow in the test section remains equal to the reservoir pressure as a consequence of Bernoulli's equation. This situation lends itself to simple aerodynamic measurements. For purposes of comparison, pressure values measured on models, in the free stream, or on walls are made dimensionless by division by p_0.

Although we can solve many important problems using the ideal-flow concept and ignoring viscosity, we must now abandon this simplification to arrive at a complete understanding of real flows. This brings us to the last—and probably most difficult—stage of our exploration of fluid dynamics before we turn to aerodynamics. We will now look at situations where internal friction, the resistance of fluid motion to deformation, cannot be disregarded.

5.3 Viscosity and Turbulence

In Chapter 3 we listed viscosity simply as a property of fluids. We saw that viscosity is a concept for which we have an intuitive appreciation. Honey is more viscous than water, which in turn is more viscous than air. Improving upon these *qualitative* observations in order to study viscous effects *quantitatively,** we envision, for example, a flow in a narrow gap—or channel—between two plates, as shown in Figure 5.8. In reality, it is more practical for the study of the behavior of viscous fluids to use two concentric cylinders with a fluid-filled gap that is much narrower than the radius of either cylinder. One cylinder is held fixed, while the other turns at a known rate. This arrangement permits the direct measurement of viscosity by a determination of the torque exerted on the fixed cylinder. The torque is here defined as a turning moment (see Chapter 9.1) of the forces acting on the cylinder. Common sense tells us that it is tougher to turn a shaft at a given rate in sticky oil than in air. But to return to our thought experiment,† we now fix plate B, while plate A is moved at a constant speed, U, by a force, F. *It is observed in nature that the fluid directly adjacent to a moving or a fixed surface adheres to it.* Thus

---

*Again, qualitative descriptions give a general idea about how things behave or work. An engineer must proceed beyond such initial impressions and find specific numerical answers to a problem. After designing a new wing that he thinks will have improved lift, an engineer must perform calculations and experiments that give actual values of the lift.

†A thought experiment is one that is never carried out! We think about the steps of a process and make sure that we do not violate a law of nature. Then we study the outcome and learn from it without actually setting up an experiment.

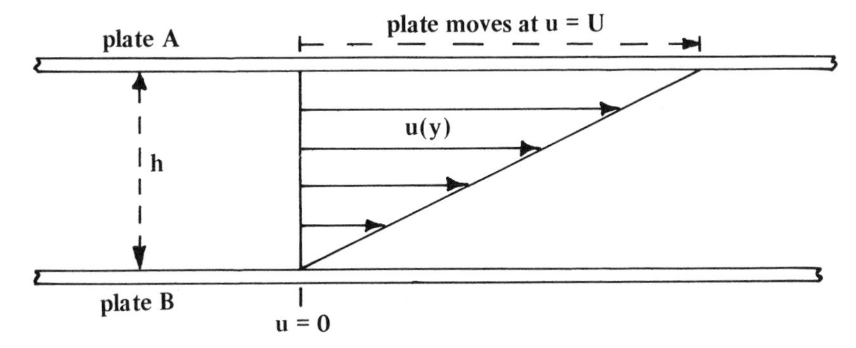

Figure 5.8. Viscous flow in a narrow *gap; an experimental demonstration of the effects of viscosity. Plate A is moved to the right at a constant speed* U *by the application of a force. Plate B is fixed, and the fluid adheres to both plates. The velocity profile* u(y) *is linear; it follows a straight line.*

the fluid in the gap is no longer ideal; it sticks to both plates, and internal friction dominates the distribution of flow speed as shown. Consider a knife pulled out of a jar of honey. The outer edge of the layer of honey moves downward. Closer to the knife the motion is retarded (this is the action of viscosity), and on the knife itself the honey appears glued to it. If you look down the side of a moving ship, you notice that the water right next to the side is dragged along with it. The same is true for air next to a moving airplane. We recall that in our ideal flow without internal friction, the fluid slips by at full speed, ignoring the presence of the surface. We will later have to reconcile the two views of viscous and inviscid fluid motion to arrive at the full view of real flows.

In Figure 5.8 we encounter for the first of many times a *velocity profile,* here a particularly simple one. The parallel arrows represent by their lengths the flow speeds of layers of fluid that are moving in the same direction. The profile itself gives flow speed as a function of distance from the fixed plate B. The coordinate y is counted at a right angle to the flow, with the origin at plate B. The local flow speed varies from zero at plate B to U at plate A. Experiments on flow in narrow gaps have shown that the flow speed varies linearly, or on a straight line, as in Figure 5.8. With this velocity distribution in the gap we have

$$u = U \cdot (y/h),$$

an equation that fulfills the *boundary conditions* of adherence of the fluid to the two plates. Note from the formula that $u = 0$ at $y = 0$, and $u = U$ at $y = h$. Surprisingly, this simple velocity law is valid in narrow gaps for fluids as diverse as oil, water, and air.

We can visualize the layers of fluid as sheets stacked on top of each other, with friction acting between them. A simple demonstration of this effect can be produced with a telephone directory. Move the cover sideways to distort the book. There is friction between every pair of adjacent pages: a sum of forces due to friction is accumulated from cover to cover. Many separate layers of pages make up this effect. The model of such layers, or *laminae* in Latin, suggested the expression *laminar flow* for fluid motion dominated by viscosity. Internal forces resist the motion of the upper plate. We define this *shearing force,* or *shear stress,* by τ (tau) = F/A, with F representing the force exerted on the upper plate of area A. This force can also be viewed as the force needed to overcome the internal resistance of the fluid due to viscosity. The shear stress has the dimension of pressure (recall Chapter 3.1). Take a piece of wood and move it on a flat surface; the force you need to overcome the friction between the wood block and the surface is the same as the shearing force in our fluid layer. This force is proportional to the slope of the straight-line velocity distribution of Figure 5.8—that is, $\tau = U/h$. It is apparent from Figure 5.8 that moving plate A at higher speed U, the straight velocity profile will tilt to the right, and more shearing force will be needed as U goes up. The proportionality constant in this relation was defined by Newton as the *dynamic viscosity,* μ (mu). Thus we have

$$\tau = F/A = \mu \cdot (U/h),$$

an expression called *Newton's law of shear.* The relation tells us that the shear force increases with increasing viscosity at a fixed velocity distribution or shear rate (U/h). In turn, at fixed viscosity, the shear force τ increases at higher rates of shear. With this step we have turned a proportionality into an equation. A proportional relation tells us that the speed of an automobile depends on power, with higher power producing higher speed. But only an equation permits us to compute the power-speed relation exactly. In Newton's law of shear we can use viscosity data such as those of Table 3.1 and actually calculate the shear force for certain velocity profiles.

Situations like that shown in Figure 5.8 occur in many applications. For example, a turning shaft in a bearing is lubricated by an oil film whose motion obeys the shear law. (The subfield of fluid mechanics dealing with lubrication has the strange name of tribology.) Bearings can also be lubricated by air films; such bearings take advantage of the low viscosity of air versus oil in order to reduce friction. Remembering the experimental technique of measuring viscosity in the gap between a rotating and a fixed cylinder (a bearing), we now understand that at lower viscosity the torque—the turning moment—will drop because of smaller internal fluid resistance.

It is impractical to keep the dimensions of more complicated properties continually in mind. However, they can be readily derived from the equations in which they appear. For example, the dimensions of dynamic viscosity are obtained from Newton's law of shear written to express the dynamic viscosity

$$\mu = (\tau h)/U,$$

by recalling that shear is given by force per unit area. We find for the dimension of viscosity $(FT)/L^2$. The corresponding SI units are $(Ns)/m^2$; conversion factors from other more traditional units are listed in Table A3.3. It is difficult to visualize the physical meaning of the dimensions and units of the viscosity. The actual values shown in Table 3.1 demonstrate the remarkably large range of values of this property—for example, water is fifty-five times as viscous as air. The table lists values measured at room temperature. At higher temperatures, the viscosity of liquids *decreases*—they become more "slippery"—while that of gases *increases*. This seemingly contradictory behavior can only be explained on the molecular level. The effect of temperature on the viscosity of liquids is well known to us; it is illustrated by the use of engine oils of different viscosities for summer and winter driving. Even simpler, think of butter at refrigerator or at room temperature.

Mercury and water have roughly the same viscosity. However, their densities differ by a factor of thirteen. Their weights also differ substantially, a fact that becomes strikingly apparent when one lifts a bottle of mercury. Thus we can see that viscosity and density are not directly related. Yet a combination of the two is a useful measure of fluid properties. Dividing viscosity by density yields the ratio ν (nu) = μ/ρ, which is called the ***kinematic viscosity*** (see Table 3.1) to distinguish it from the dynamic viscosity. For incompressible flows—flows where the density remains constant, such as those of liquids and low-speed gases—the temperature also remains constant. At fixed temperature the dynamic viscosity has a fixed value (see Table 3.1), and in turn the value of the kinematic viscosity remains constant. This fact will later turn out to be helpful to simplify model tests and aid our understanding of aerodynamic parameters such as the well-known drag coefficient (see Chapter 6 and Appendix 2). Finally, it is important to observe that many fluids of technical importance, such as water, air, and various oils, obey the simple law of viscosity formulated by Newton. Therefore fluids for which the shearing force is a linear or straight-line function of the *strain,* or shearing rate (given by the slope or angle of the U/h line), are called Newtonian fluids. High shearing rates imply high values of U/h (or high velocities of plate A in Figure 5.8). The U/h line is tilted to smaller acute angles at plate B.

We have employed many of the ideas and laws—gravitational attraction, force, viscosity, and so forth—first coherently described by Isaac Newton (1642–1727). Newton was born in the year in which

Galileo Galilei died. An important figure in his own right, Galileo—who seems always to be called by his first name—was one of the important predecessors of Newton who initiated quantified science based on experimentation. Galileo studied the laws of free fall, the pendulum, and many other mechanical problems. With a simple telescope of Dutch design, he discovered the four major moons of Jupiter and Saturn's ring. He joined the early proponents of the view that the earth revolves about the sun, an act that incurred the disfavor of the Catholic church, whose inquisitional court silenced him.

Newton's studies at Cambridge University led to his graduation in 1665 without particular distinction. He then went to his mother's farm, since London was in the midst of one of its recurrent outbreaks of plague. In the short period of 1664 to 1666—he was in his early twenties—Newton worked out the law of universal gravitation, laid out the basis of calculus, and discovered that white light, if refracted through a prism, displayed a spectrum, a sequence of colors as seen in a rainbow. In 1669 he became a professor at Cambridge University, a post he held until 1701, doing experiments involving sound, viscosity, and other phenomena related to our interests. Much of his work was published, at the urging of others, in 1687 in a book whose long Latin title is usually abbreviated *Principia*. Newton brought clarity to many ideas of his predecessors and contemporaries and added important original thoughts of his own. In the process of finding solutions to many problems in fields such as mechanics and sound, he developed independently a new form of mathematics, the calculus, about the same time as the German mathematician and philosopher Gottfried von Leibniz (1646–1716). The latter also calculated the forces exerted by air to support an object. This work was published in a paper with the title "Aeronautik," the first mention of the now-familiar term aeronautics. Considering his many contributions, Newton must be counted among the greatest minds in the history of physics.

But to return to viscous flow. Internal friction in viscous flow causes the total energy of the fluid to *decrease*. The friction, just like the friction of a poorly greased wheel bearing, requires energy that is lost to motion. This decrease of energy is manifested by a decrease in the stagnation pressure. Applying Bernoulli's equation to two locations in a viscous pipe flow, we observe that the sum of dynamic and static pressure—the stagnation pressure—decreases. This decrease represents the amount of the energy loss owing to internal friction. Thus Bernoulli's equation describing ideal flow permits the determination of energy loss in viscous flow, another most useful property of this important theorem. You will ask how this observation fits the law of conservation of energy. Here we have to understand temperature changes in the fluid, an effect absent in ideal flow. The internal friction due to viscosity causes what is called *dissipation,* an irreversible—a one-way—transformation of mechanical energy into heat, which is reflected in a slight temperature increase in the flow. Returning to our bearing, we find that the friction heats the bearing; consequently energy is lost to heat in this mechanical system. The total energy—now mechanical energy (or work) plus heat—remains constant. This situation demonstrates the first law of thermodynamics, a fundamental law that applies to the conservation of the sum of mechanical and heat energy. As we have seen, the same conservation principle applies to all forms of energy.

We have now looked at some aspects of viscous flows involving simple laminar motion. However, most flows in nature and technology are of a radically different character. Fully viscous (or laminar) flow in water and air is restricted to low speed. At higher speeds the orderly motion of laminar flow shown in Figure 5.8 ceases to exist. Observation reveals an irregular, swirling, random fluid movement, a breakdown of the regular layers. Correspondingly, the local speed at a given point in steady flow, whose average properties do not change with time, is found in the disorderly flow to fluctuate or jiggle about a mean value in all three directions. Such flows are called *turbulent*. The shear rates in turbulent flow, which are much higher than those in laminar motion, can only be predicted empirically at our current level of understanding. The swirling, irregular motion produces a higher loss of energy than the more orderly friction between adjacent layers of fluid. Turbulent flows are all around us. They were observed by Leonardo da Vinci, who noted the vortices in the wake of boulders in a river and the turbulent breaking of waves. The motion of clouds, the rapid mixing of cream in a cup of coffee, swirling leaves or snowflakes in the wind, the buffeting of a car by a passing truck, the surface of a creek or the surface of the sun, the irregular structure of a spiral nebula, the flames of a forest fire, the wake of a speedboat, the air pushing an airplane about—all are examples of the effects of turbulent flows. The words "turbulent" and "turbulence" have even entered our daily language to describe disturbances, commotion, agitation, violence, and the like.

The two states of laminar and turbulent flow are both seen in the vertically rising smoke of a cigarette. Close to the cigarette a straight filament is observed. At some point in the upward motion, this quiet flow turns into turbulent flow. Irregular motion appears; the smoke spreads and looks like a cloud in rapid motion or the billowing smoke from a smokestack. The process of change between the two modes of flow is called *transition*.

To improve our understanding of the two fundamentally different modes of flow, laminar and turbulent, it is useful to take a look at the earliest research in this field. The first systematic investigation of the two kinds of motion involved pipe flow. The German hydraulic engineer Gotthilf Heinrich Ludwig Hagen (1797–1884) tested the flow in small-diameter tubes and determined the pressure drop with distance in the tube. This pressure drop is a measure of the dissipation, or energy loss by friction, due to viscosity in the pipe,* which we noted a moment ago when we described the application of Bernoulli's equation to a measurement of it. Hagen studied mostly what we now call laminar pipe flow; however, he was the first to observe that, under certain conditions—for example, when he used water at higher temperatures (and consequently at lower viscosity) or in larger tubes—the flow's character was decidedly altered. By switching to glass tubes and adding sawdust to the water to make the flow patterns visible, he was able to observe a whirling, irregular motion, discovering in 1839 what we now call turbulent pipe flow.

Concurrently, and unaware of Hagen's work, a French physician, Jean Louis Poiseuille (1799–1869), became interested in flow in small pipes with applications to experimental physiology. The flow of blood readily comes to mind as posing problems in biological fluid dynamics. Poiseuille's careful experiments, which concerned laminar pipe flow only, led later to the exact mathematical derivation of the velocity profile of viscous flow in pipes. The distribution of velocity as a function of pipe radius was found to be parabolic,† with $u = 0$ at the pipe wall and with the maximum speed on the centerline. The beautiful simplicity of what represents an analytical—that is, an exact—solution of the complete equations of motion including viscosity emerges. This type of solution of the equations—provided the properties of a fluid are known—describes the flow exactly without resorting to experimentally determined correction factors. Why do we discuss laminar flow at all in view of our previous statement that nearly all flows in aerodynamics or other technical applications are turbulent? For one thing, in Chapter 10, where we discuss plans for the improvement of the performance of modern aircraft, we shall find that attempts to achieve laminar boundary layers are an important project; therefore we must learn about laminar flows to understand where they can help us. Moreover, a rounded picture of fluid dynamics includes laminar flow because of the intrinsic interest of this type of motion.

The definitive understanding of the two modes of flow, and in particular of the transition from laminar to turbulent motion, is also associated with pipe flow. The Englishman Osborne Reynolds (1842–

*Pumping stations are required at certain intervals along a pipeline to overcome the pressure drop and boost the pressure to a value sufficiently high to keep fluids such as oil, water, or natural gas moving.

†Parabolic means here that the velocity profile from wall to wall in the pipe describes a parabola.

1912), who worked in many areas of physics and engineering, published in 1883 a paper entitled "An Experimental Investigation of the Circumstances Which Determine Whether the Motion of Water Shall Be Direct or Sinuous, and of the Law of Resistance in Parallel Channels." We now call direct and sinuous motions laminar and turbulent, respectively. Reynolds's original results for pipe flow are qualitatively identical to those later observed for many other types of flow, near walls, in vertically rising smoke columns, on airplane wings, and elsewhere.

Reynolds used a glass apparatus in which water flowed from a reservoir into a pipe whose center streamline was marked by dye. The inlet from the reservoir to the pipe was carefully shaped to avoid disturbances such as vortices arising at sharp corners. In the pipe the dye filament remains smooth for some distance. This smooth (or direct) motion of the laminar pipe flow changes at some point farther downstream—the transition—to a wavy (or sinuous) motion, and next it becomes turbulent. The dye is quickly dispersed from wall to wall. In turbulent flow, the velocity profile in the pipe is drastically altered from the parabolic one of laminar flow. It is more pluglike: the flow moves at roughly the same speed all across the pipe, but near the wall there is a narrow region where the speed quickly reaches zero.

The two basic types of flow with their very different velocity distributions occur in many other contexts. They dominate the flow near solid surfaces, where the speed must be reduced to zero. As mentioned before, the shear forces in turbulent flow are much higher than those of laminar motion. Correspondingly, a higher pressure loss with distance in the pipe is associated with the churning motion. The wall friction, or *skin friction,* is also higher in turbulent flow, an effect with which we must deal when we discuss the aerodynamics of airplanes, whose shape is designed for a low air drag but whose large surface area is exposed to surface friction that adds to the overall drag (see Chapter 7.3).

5.4 The Boundary Layer

Let us now turn to viscous effects in flows other than those in a pipe. Remember that fluids in motion adhere to a surface. The surface must be more or less aligned with the flow direction; shapes such as blunt rear sections of bodies produce separation, a complete breakaway of the fluid from the object such as is observed in the wake of a bullet, at the stern of a rowboat, or behind a boulder in a creek. The effect of the presence of walls is transmitted to the flow by the retardation of succeeding layers of fluid, which is finally brought to a halt at the walls by the mechanism of internal shear. Ideal fluids—fluids without viscosity—would simply slip by at full speed. With the flow speed of a real fluid approaching zero at a wall—it "sticks" to the wall—the change of flow speed near the surface, its *gradient,* becomes large. This essential fact remains, no matter how small the viscosity of a fluid may be; the shear in the fluid near a wall must be taken into account.

The understanding of these deceptively simple ideas is very recent; indeed, it came early in this century. A flow model accounting for the effects of shear in low-viscosity fluids was proposed in 1904 before the Third International Congress of Mathematicians by the German engineer Ludwig Prandtl (1875–1953). As Prandtl described his work, "I have set myself the task of investigating systematically the motion of a fluid of which the internal resistance is supposed to be so small that it can be neglected wherever great velocity differences . . . do not exist." He further suggested that "the investigation of a particular flow phenomenon is divided into two interdependent parts: there is on the one hand the *free fluid,* which can be treated as inviscid . . . and on the other hand the transition layers at the fixed boundaries." These layers are confined to a narrow zone, now called the *boundary layer* as translated from the German *Grenzschicht,* while the word *transition* is now reserved for the change from laminar to turbulent flow that we examined above. In the region adjoining a solid surface, the flow speed changes from that of the outer or ideal fluid, whose internal friction or viscosity can indeed be ignored in the case of water or air, to zero at the surface.

The introduction of the ingenious boundary-layer concept by Prandtl early in this century marks the beginning of modern fluid mechanics; indeed, it revolutionized our understanding of the field. The great importance of Prandtl's contribution went largely unnoticed by the mathematicians present at his original 1904 lecture, yet Felix Klein of the University of Göttingen, a leading figure in applied mathematics, was so impressed by the proposed concept that he arranged for Prandtl to direct a research institute at Göttingen. Prandtl's unusual combination of physical insight and mathematical ability soon turned his small research institute into a world-renowned laboratory.

Modern aerodynamics is strongly dependent on an ever-improving understanding of boundary layers, in particular those that are turbulent, which continue to be a centerpiece of current research. Although great progress had been made in the nineteenth century based on the mathematical understanding of ideal flow, serious discrepancies with respect to experimentation remained. The equations of motion for inviscid flow in three dimensions such as the flow about a sphere had been solved, and streamline patterns about even more complicated shapes had been computed. Oddly, however, many of the results did not agree with observation at all. Typically—and we shall soon see why—it was calculated that a sphere has no aerodynamic drag. Since anyone who ever threw a ball knows that this is nonsense, something must have been drastically wrong with the theory.

In parallel, as we have already seen, a substantial body of empirical knowledge had been compiled through the centuries for real (versus ideal) viscous flow. Prandtl's boundary-layer concept joined together the two diverse strands of knowledge in fluid dynamics. Ideal flow dominates large domains in a flow field about an airplane. But close to the wings, the fuselage, and the control surfaces, friction can never be neglected. Near the surface in the boundary layer, the flow speed decreases to zero. Air has a very low viscosity, but the rapid velocity change in the boundary layer still produces a large *shear gradient* that leads to friction forces. Other important effects such as flow separation from the surface, with which we shall deal later, are indirectly caused by the boundary layer.

The boundary layer is most readily visualized by sketching the steady flow past a flat plate in the wind-tunnel picture shown in Figure 5.9. The outer "edge" of the layer, or its thickness, δ (delta), is defined as the distance from the surface to where the actual flow speed, u, is nearly equal to that of the undisturbed *free stream*, or *outer* flow speed U. At that point the presence of the surface is not reflected in the flow. Like the flow in a narrow gap (see Figure 5.8), flow speed measured at a right angle to the plate, $u(y)$, varies from $u = 0$ on the plate (i.e., $y = 0$), to $u = U$ at $y = \delta$. Like the flow in a pipe, *the boundary layer may be laminar or turbulent.* The boundary layer grows with distance from the leading edge, the tip, of the plate. In the initial stages of growth, the shear is dominated by viscosity; the boundary layer is laminar. The transition between the two states of flow, laminar and turbulent, occurs on smooth plates under relatively well-defined conditions which we will enumerate later. This fact again reminds us of pipe flow, or of the rising flow from a cigarette or a

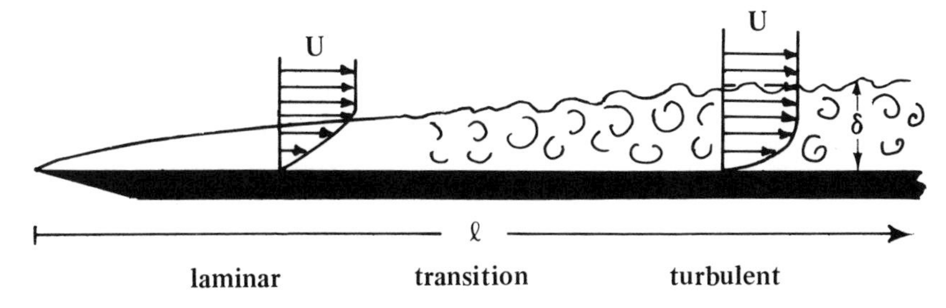

Figure 5.9. The boundary layer on a flat plate exposed to a uniform external flow of speed U *from left to right. Note the velocity profiles of laminar and turbulent boundary-layer flows, with the transition between the two states. For both types of motion the flow speed is zero at the plate surface.*

smokestack. However, in nearly all practical applications, ranging from rowboats to aircraft, on propellers and in jet turbines, transition to turbulence occurs near the leading edge or forward point of the device. Consequently, just as was the case before in pipe flow, we usually deal in nature and technology with turbulent boundary layers.

This fact is somewhat unfortunate, since the structure of laminar boundary layers—just like the laminar pipe flow—can be computed exactly from the flow equations for viscous fluids. The turbulent boundary layer, on the other hand, although tested for many years in just about every conceivable flow situation, is computed largely from empirical information. Furthermore, turbulent boundary layers are thicker than laminar boundary layers (see Chapter 7.3). The velocity profile of Figure 5.9 demonstrates schematically the rapid drop of flow speed near the wall in a turbulent boundary layer leading to a higher wall friction than that of the laminar layer. Here we remember our formula for shear stress, an expression that also represents qualitatively the behavior of turbulent boundary layers.

The events in the transitional region between the two types of flow are even more difficult to understand. Like the phenomenon of turbulence itself, it is still one of the major research areas in fluid dynamics. Another difficulty lies in an accurate prediction of the location of the change from laminar to turbulent flow in different situations. Better understanding of transition is essential for improvements in the aerodynamic performance of aircraft. But today turbulence itself in all its guises is still largely an enigma. For this reason research in the field is often directed to an improved understanding of the basic physics of the problem. Advanced experimental equipment and new methods to visualize flows like free jets, wakes behind bodies, and boundary layers have improved our understanding of this phenomenon. Help has come from different sources, since aside from our field, it is of great interest in astrophysics, meteorology, and physics as a basic random (i.e., irregular, unpredictable in detail) process. These remarks must not discourage us. As said before, a great body of semi-empirical work on the effects of turbulence has contributed to safe flying, improved steam turbines, more efficient combustion in automobile engines, and other technical applications.

The turbulent boundary layer is responsible for most of the hissing noise one hears in a jet airplane. The noise comes from irregular pressure pulses beating on the cabin wall. The water boundary layer on sailboat hulls, the air boundary layer on the wings of birds and around automobiles, and the atmospheric boundary layer on the surface of the earth are all turbulent. In fact, the boundary layer on the earth has a height of the order of about 1,500 m (4,900 ft). The mixing process in the lower atmosphere of water vapor or industrial emissions, the plume of a smokestack, and rising heated surface air are all examples of turbulent motion near the earth's surface. Turbulent motion can be observed when snowflakes or grains of sand swirl in the wind and when waves are whipped up on bodies of water.

The early part of this century, a period coinciding with the realization of airplane flight, saw the birth of modern fluid dynamics arising from the merger of ideal-flow theory, with all its splendid mathematical solutions of flow problems, and the boundary-layer concept. The idea of the boundary layer excludes the use of flow theory which ignores viscosity close to surfaces. In this narrow region viscosity and turbulence dominate. Take a streamlined object such as the airfoil section shown in Figure 5.3. Here the streamlines of the flow are marked. In addition, however, the wing is also shrouded in a thin boundary layer in which shear and friction cause a drag or resistance to the motion of the wing. Most importantly, the phenomenon of flow separation (see Chapter 7) is induced by the presence of the boundary layer. This literal break of a flow leaving the surface of an object is found in many places, ranging from the trailing edge of wings of aircraft and birds, to the tops of automobiles, to blunt objects such as water towers, telephone poles, and buildings.

The thin boundary layer near the leading edge of the wing is not discernible in the photograph of Figure 5.3. However, turbulent motion can be seen at the trailing edge. Aerodynamics requires a broad understanding of viscous flows, flows governed by turbulence, and flows where viscosity can be neglected, and it requires the insight and ability to piece together the characteristic features of the various types of flow in the solution of a given problem. Finally a coherent picture of the aerodynamics of an airplane emerges, and the forces that act on the craft can be understood.

Chapter Six

Turning to Aerodynamics

The success of any physical investigation depends on the judicious selection of what is to be observed as of primary importance, combined with a voluntary abstraction of the mind from those features which, however attractive they may appear, we are not yet sufficiently advanced in science to investigate with profit.

James Clerk Maxwell (1831–79)

6.1 How Do We Test Models of Airplanes?

We have explored the early days of flight and other aspects of aviation. We have also examined properties of fluids and the behavior of fluids at rest and in motion—the field of fluid mechanics. The atmosphere, the medium in which airplanes fly, has also been discussed. With this background, we can now turn to aerodynamics, the study of the forces acting on all objects moving in a fluid. Natural and man-made objects are included; from fish to submarines, from birds to aircraft, all are subject to the same laws of aerodynamics.

But how do we handle experimental results obtained from models or full-scale aircraft, how do we test and improve new designs, and how do we arrange the accumulated information? A scheme must be found to master the abundance of experimental and theoretical results and to facilitate comparisons by putting the results on a comparable footing. We have to advance from specific findings to more general functional relationships valid in a wide range of applications. Assume that we have measured in a wind tunnel the aerodynamic drag in units of force on a set of scale models of different airplanes. A scale model of a prototype is (externally) a smaller version of the real thing; the two look exactly alike. It is said that the model has *geometric similarity.* How do we apply our drag measurements to the full-size airplane under design? The route to this goal lies in finding a *dimensionless* expression that accommodates a wider use of the initial set of experimental results.* Our experiments should yield general information, provided they were well thought out.

Enough of these abstractions. Aerodynamic development still depends largely on the testing of models in a wide array of testing facilities, of which the wind tunnel is still the foremost one, although computers play an increasingly important role. Various other testing methods are described in Appendix 2.1, but we shall concentrate on the wind tunnel. We start with the small scale model of an airplane or a component of the airplane such as a wing. We expose the model to the airflow in the wind tunnel. The flow field about the model (see Figure 5.3) is determined by the size of the model, the properties of the air (such as density), and the wind speed. How do we apply the measurements of lift and drag taken from the model in the wind tunnel to the full-size airplane? To accomplish this a *similarity parameter* is needed that tells us under what conditions the flows about model and prototype are similar. Here similar means that the streamline pictures look identical; photographs taken of both, and enlarged or reduced to the same size, cannot be distinguished.

The sketch of the pioneering Cal Tech wind tunnel shown in Figure 6.1 is characteristic of this type of testing equipment. Similar facilities exist in many university, government, and industrial laboratories around the world. The tunnel shown operates at wind velocities substantially below the speed of sound in the range where air can be regarded as incompressible. Therefore it is called a *low-speed wind tunnel.* A *uniform* airflow (a flow that is everywhere at constant speed) is blown against the model mounted in the *test section,* the place where the model and its instrumentation are located. Above the test section is a balance room. The rods and wires with which the model (not shown in Figure 6.1) is suspended are led outside to this room. The suspension system is connected to a beam balance with which the aerodynamic forces such as drag and lift (see Figure 1.1) can be determined. The test section of a huge low-speed wind tunnel oper-

*I again assume that the reader is familiar with Appendix 2.2.

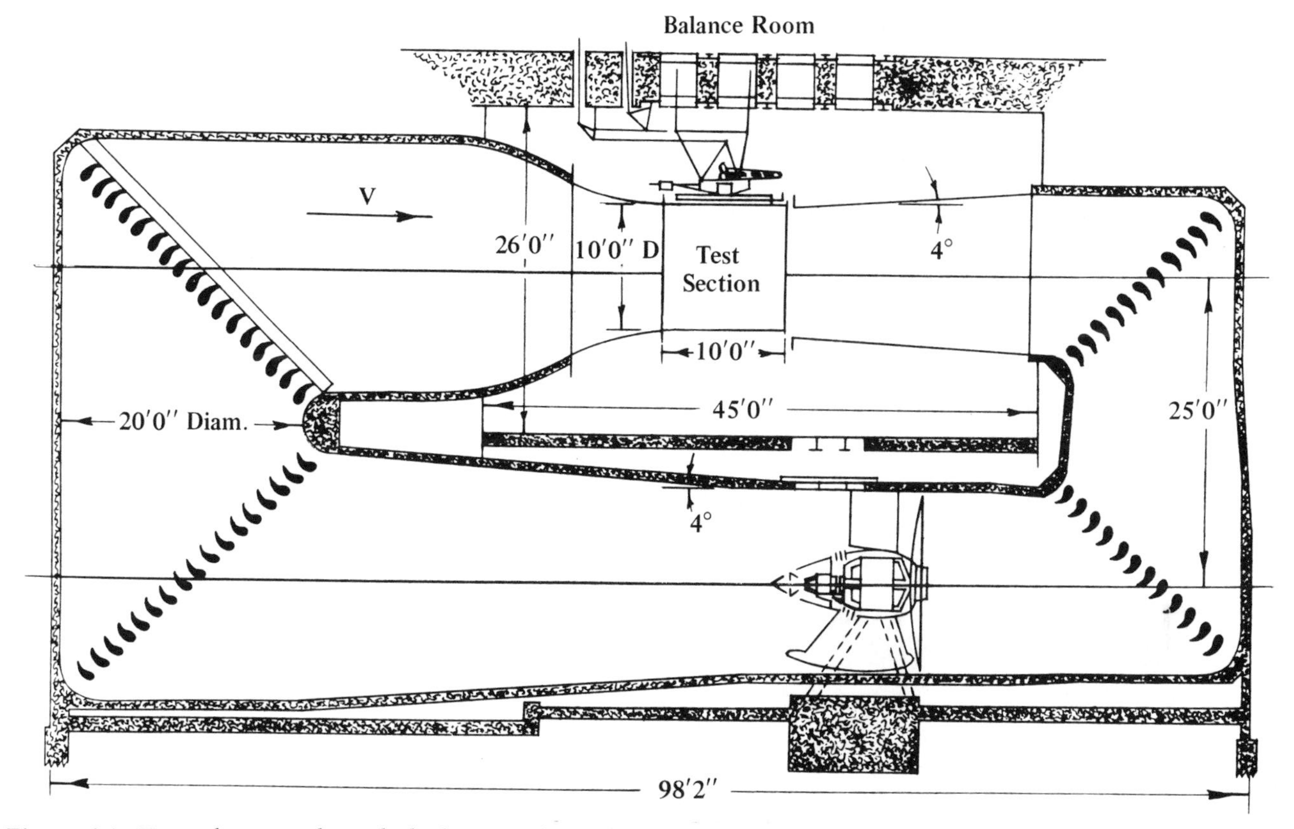

Figure 6.1. Vertical section through the by now classical 10-ft low-speed wind tunnel and six-component suspension system of the Guggenheim Aeronautical Laboratory of the California Institute of Technology, Pasadena, California.

ated by NASA is shown in Figure 6.2. The large model installed in this test section is that of a supersonic airplane, an aircraft that flies faster than the speed of sound. Even supersonic airplanes must start and land at low speeds, so that their aerodynamic characteristics must be tested in low-speed wind tunnels. Most wind tunnels that are not as large as the NASA tunnel of Figure 6.2 operate with closed circuits like that shown in Figure 6.1. A large fan starts the air moving, and then the air is carefully guided around the corners and led via straighteners and screens to the test section. The object is to reduce the turbulence in the flow to a minimum in order to duplicate conditions of an airplane moving in still air.

We have already noted that it is convenient to address aerodynamic problems in a wind tunnel, with the model fixed with respect to the observer. Indeed, all figures in this and similar books depict such a setup, for obvious reasons. This is a safe procedure, since iden-

Figure 6.2. Model of a supersonic aircraft mounted in the test section of the 30-by-60-ft, low-speed wind tunnel at the NASA Langley Research Center. The supports are connected to devices by which drag, lift, and other forces (or moments) can be measured.

tical conditions prevail in free flight in calm air and in the wind tunnel, provided a uniform flow at low turbulence is achieved in the test section. Well below the speed of sound, the flow is *incompressible,* its density remaining constant. Changes of pressure in the airflow about the model, or about the real airplane, are moderate and do not substantially alter the air density. Since, as we learned before, water is incompressible, the equations used in aerodynamics apply equally to water and air. Consequently, the similarity conditions for the low-speed range govern the motion of a trout as well as a commuter airplane; indeed, both have been tested in wind tunnels.

To consider similarity in an incompressible flow, we recall from Chapter 5 that inertial and viscous forces dominate; they must both be considered in model testing (see Appendix 2.3). The interplay of these two effects shapes the fluid motion about an object. Recall that we are attempting to obtain identical (similar) flow pictures about the prototype and its scale model. First we decide on model size, say one-tenth the size of the original airplane. Therefore a single characteristic length dimension, such as the chord of a wing, designated by l, enters considerations of similarity. The ratio of that length from the model to the real thing indicates the scale, in our example 1:10.

Flow fields are similar provided the ratios of the inertial to the viscous forces acting on a given fluid element at corresponding locations in the flow fields are identical. This is the scientific statement of the basis of similarity. In simpler terms, if indeed the force ratios are identical everywhere in the flow about the airplane and its scale model, the entire flows about both—that is, the so-called flow fields—will look identical. Marking the streamlines with smoke (see Figure 5.3), we would not be able to tell them apart except by the difference in size. Note that we stipulate a *ratio* of the important forces that control the motion, not their absolute values. Clearly the lift of an airplane—a force—will be larger than that of its model in the wind tunnel, and for model testing we must find a way to eliminate absolute values in order to compare data. We further assume that the geometrically similar objects are both surrounded by fluid; they are fully immersed in it. Our thoughts apply to a fish or a submarine, and a bird or an airplane, but not to a ship cruising on the surface of the ocean.

The ratio of the inertial to the viscous forces can be worked out by relatively simple physical reasoning. This is done in Appendix 2.3 (albeit without any higher mathematics) for those readers who opt for a deeper understanding of similarity. However, that discussion can be omitted, and only its result will be shown here. The desired dimensionless *similarity parameter* is given by

$$Re = \frac{Ul\rho}{\mu} = \frac{Ul}{\upsilon}.$$

This expression is called the *Reynolds number*, designated by Re, in honor of Osborne Reynolds, whom we encountered in Chapter 5. (The other symbols in the equation appear in the previous chapters and in Table A2.1.) By writing out the dimensions of the parameters in the Reynolds number, one can see that the expression is indeed dimensionless. Regardless of the absolute values of the speed of motion, the density, and so forth, if the dimensionless Reynolds numbers of model and prototype are identical, the flow fields about both will be identical. We further note the kinematic viscosity, ν, in the third term of the equation, a property value listed in Table 3.1. The kinematic viscosity is the ratio of the dynamic viscosity to the density. This ratio simplifies calculations in low-speed or incompressible flow. Here the kinematic viscosity is constant because the dynamic viscosity and the density are both constant: thermodynamics and temperature effects do not enter.

The formula for the Reynolds number tells us that when we are testing a model, we can change any of the air properties, or wind speed and model length, to duplicate the Reynolds number of the full-size flying object. Say that we have a 1:10-scale model. All we have to do in the wind tunnel to obtain complete similarity is blow the wind in the test section at ten times the flight speed of the airplane. We can measure drag and lift forces and determine the distribution of pressure, and—using dimensionless force coefficients, to be given in the next section—the results are directly applicable to the full-size aircraft. In this experiment we traded size for speed. But we have to watch that the wind-tunnel speed does not become so high for small models that it exceeds the range where the air remains an incompressible medium. Making larger models, aside from incurring increased costs, leads to technical problems with suspension in the high wind, and correspondingly high forces that act on the model in test sections of the kind we saw in Figure 6.2. Here the suspension system is articulated to alter the angle of the model in relation to the oncoming flow. From thermodynamics, we know that at a fixed temperature the density is directly proportional to the pressure. Twice the pressure gives twice the density. Therefore we can increase the Reynolds number by pressurizing the entire wind-tunnel circuit shown in Figure 6.1. A final trick to increase the Reynolds number—the lowering of viscosity—will be related in Chapter 10 in conjunction with wind tunnels operating near or at the speed of sound, so-called *transonic* wind tunnels.

Because of the importance of Reynolds numbers for testing and for sorting out types of flow (laminar versus turbulent or fully viscous flows such as the Stokes flow about the sphere discussed in Chapter 7.2) as well as in formulating functional relationships between other aerodynamic parameters, we again refer to the detailed discussion of the topic in Appendix 2.3. The method demonstrated there is characteristic of the way in which engineers approach complex problems, an approach that sets engineering thought and methods of reasoning apart from those of physics. Complex relations—here the variation of speed, length, density, and viscosity—must be considered simultaneously. The physicist tries to study one effect after another by reducing the variables in an experiment to one at a time. To provide values such as those shown in Table 3.1, for example, the physicist studies the viscosity as a function of temperature. This can be measured at one pressure for one material, eliminating all other possible variables. In aerodynamics, as in other engineering disciplines, there is no getting away from greater complexity: an airplane must fly, and many factors affect its performance simultaneously. To meet this need, engineers have devised many dimensionless similarity parameters dealing with specific combinations of effects. We have looked at one; more of them will come up later. It has been the special genius of engineers throughout history to develop original methods; the oversimplification that goes something like "scientists find out about things and engineers make them work" overlooks the brilliant contributions to fundamental knowledge provided by engineers.

What is the range of Reynolds numbers of moving objects? At sea level, the Reynolds number of an automobile of a given length is simply a linear function of the speed; twice the speed, twice the Reynolds number. But the kinematic viscosity is a function of temperature and density; consequently, it varies with altitude in the atmosphere. It can be computed by finding density and temperature at a certain altitude in Table 4.1 and noting the viscosity of air for that temperature in a standard table. With a known flight speed and length, the Reynolds number can then be computed for the many things that fly through the air. The result is an enormous range of that parameter, as shown in Figure 6.3. Logarithmic scales (see Appendix 1) en-

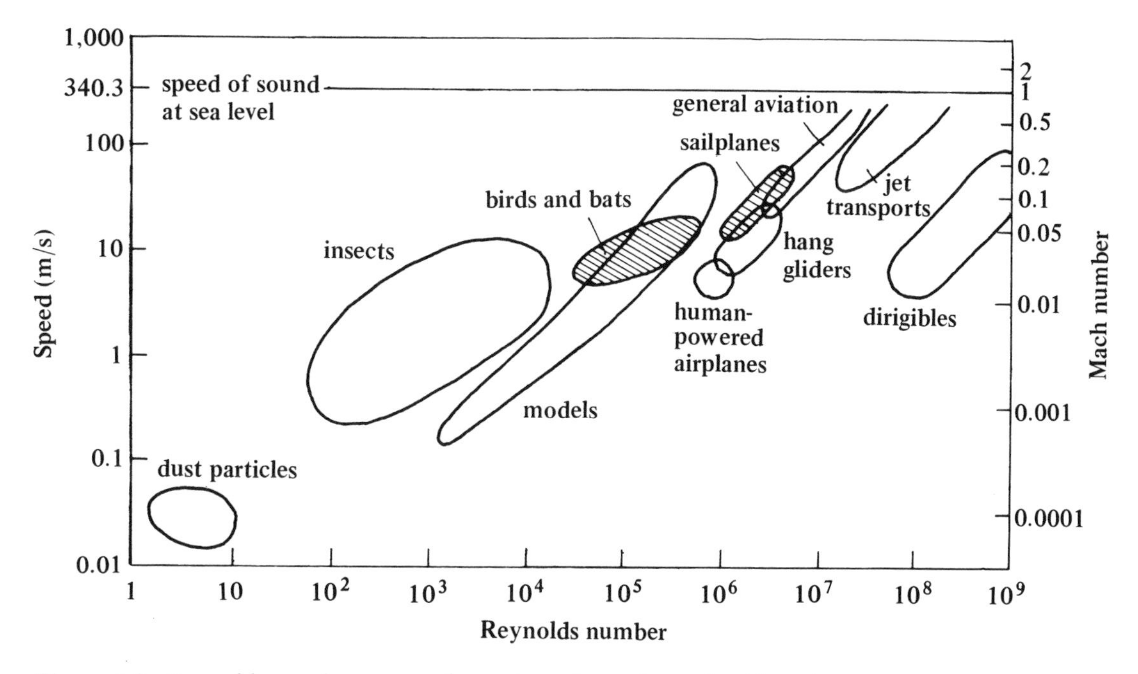

Figure 6.3. Reynolds-number range of a variety of aerodynamic objects in nature and technology. Reynolds numbers are given as a function of Mach number and speed. The ranges indicated for the different types of flying objects are approximate.

able us to cover the motion of dust particles and airships or dirigibles in one graph. The plot extends on its ordinate to speeds in the range of compressible aerodynamics—speeds that approach or even exceed the speed of sound—for which the Reynolds number is not a sufficient parameter.

The *Mach number*—well known to every newspaper reader—enters as a new similarity parameter once flight speeds approach the speed of sound, as shown on the vertical scales of Figure 6.3. Looking at these scales, we recall that flight at the speed of sound is equivalent to flight at Mach number one. At sea level at the temperature of 15°C of the standard atmosphere (see Table 4.1), this speed is 1,222 km/h (759 mph). At higher altitudes the temperature decreases (see Table 4.1), and the speed of sound decreases. Depending on the accuracy with which we want to determine aerodynamic data, the Mach number may be ignored as a required similarity parameter up to flight speeds of about one-third the speed of sound. In this range the Reynolds number dominates similarity. For later use, however, we define the Mach number by

$$M = U / a,$$

which is the ratio of the flight speed, U, to the speed of sound, a. The latter is determined at the altitude of flight in the atmosphere. In the wind tunnel the speed U describes the free-stream wind speed in the test section, and the speed of sound is that computed for the test-section temperature.*

The term Mach number honors Ernst Mach (1838–1916), an Austrian physicist and philosopher. (The symbol M is analogous to the Re of Reynolds number.) Mach made major contributions to gasdynamics, the fluid dynamics of speeds near or above the speed of sound. Mach and his associate P. Salcher produced the first photographs of objects flying at supersonic speeds ($M > 1$), similar to those shown in Chapter 10.2.

6.2 Drag, Lift, and Other Coefficients

Dimensionless values of drag and lift are needed next to apply the forces determined in wind-tunnel experiments with models at the correct Reynolds number to those of the prototype. These coefficients are

*The speed of sound is proportional to the square root of the absolute temperature. This temperature is defined in Table 4.1. At lower temperatures, as noted, the speed of sound decreases. If, for example, the Concorde flies at $M = 2$ in the stratosphere, its ground speed is lower than that which would be computed for $M = 2$ at sea level, since the upper atmosphere has a lower temperature.

given in Appendix 2, and they will be applied in the next two chapters. Here we will simply state their definitions. The *drag coefficient,*

$$c_D = \frac{D}{qS},$$

and the *lift coefficient,*

$$c_L = \frac{L}{qS},$$

are dimensionless expressions obtained by dividing the drag and lift forces (see Figure 1.1), designated D and L, by the product of an area and the dynamic pressure based on the wind speed U in the test section or the flight speed of the real object. The dynamic pressure is known to us from Bernoulli's equation, where it was defined by $(\rho/2)U^2$. Obviously, consistent units must be used in this calculation. The area S in the formulas for the two coefficients—like the length in the Reynolds number—can be chosen at will. It must simply be defined in the same fashion for both the model and the full-size version. The drag coefficient for a car, a ball, or a streamlined body is commonly based on a cross-sectional area. In the case of the lift of an airplane or a wing, the area of the lifting surface is often taken. Inherent in these definitions is the fact that the actual drag and lift forces—like the coefficients—are proportional to the density of the fluid, the cross-sectional or plane area of the moving object, and the square of the flow speed. The latter two proportionalities arise from the definition of q, the dynamic pressure. This result is obtained if we multiply the coefficients by qS to get the forces. The dependency of the forces on the square of the speed was first described by Isaac Newton in his *Principia,* as we mentioned in Chapter 5.

We have now put together all the necessary ingredients to perform the calculations for transferring wind-tunnel results to the full-size airplane. From the wind-tunnel experiment we obtain the absolute values of drag and lift forces in pounds or newtons. The model's geometry defines the value of the area inserted in the calculation of the coefficients, and the dynamic pressure is given by the wind speed and density in the test section. By changing wind speed we alter the Reynolds number; we can also vary the angle of attack, α, of the model (see Figure A2.2) and obtain functional relations like C_D (Re) and C_D (Re, α). Reversing the calculations and inserting the flight conditions for the full-size craft results in the prediction of its drag and lift in absolute terms for the model. More about the physics behind these interesting relationships and their applications in aerodynamics follows in the next two chapters, which deal with aerodynamic drag and lift.

Finally, the conversion of pressure measurements on model and full-size object is important. A *pressure coefficient* is defined by

$$c_p = \frac{p - p_\infty}{q},$$

again using the dynamic pressure, q. Here p is the pressure measured on the surface of the object, and p_∞ gives the *free-stream pressure* in the wind tunnel. The latter is determined far from the model, whose presence must not affect its value. For flight, the free-stream pressure is the ambient atmospheric pressure, just as U in the dynamic pressure is the flight speed. In Appendix 2 we show that $c_p = 1$ at the stagnation point (the nose) and that the pressure coefficient may be positive or negative, depending on whether p is larger or smaller than p_∞. In discussions of the stability of airplanes, yet other coefficients arise, which we will look at in Chapter 9.1.

Chapter Seven

Aerodynamic Drag

There is no part of hydrodynamics more perplexing to the student than that which treats of the resistance of fluids. According to one school of writers a body exposed to a stream of perfect fluid would experience no resultant force at all, any augmentation of pressure on its face due to the stream being compensated by equal and opposite pressures on its rear. . . . On the other hand it is well known that in practice an obstacle does experience a force tending to carry it down stream, and of magnitude too great to be the direct effect of friction.

LORD RAYLEIGH (1842–1919)

7.1 What is Drag? Blunt Bodies

It is a long-recognized fact of experience that the motion of an object through a fluid is impeded by forces that act on the object. Whatever the exact origin of these forces resulting from the complicated interplay of body shape, the properties of the fluid, and the flow speed, a pull is exerted opposing the direction of motion. As shown in Figure 1.1, the force experienced in the direction opposite to that of an airplane's motion is called *drag*. In general, drag or aerodynamic resistance is well known to us as part of our daily experience. We feel this force fighting us when we swim, stick a hand out the window of a moving car, cycle against the wind, or ski. Any object moving through a fluid will eventually be brought to rest by the drag unless it is impelled to counter this force. The full understanding of this familiar effect is relatively recent. Lilienthal, in his glider research (1890), called all air forces "drag." Alexandre Gustave Eiffel (1832–1923), the brilliant French engineer who also did research in aerodynamics (1910–14), using his famous tower for drop tests of models and devising a new type of wind tunnel, spoke of "la résistance de l'air."

But let us look at this force from a current point of view. By rewriting the equation that defined the drag coefficient (Chapter 6.2), we have

$$D = c_D S q = c_D S\,(\rho/2)U^2.$$

This expression relates to a similar equation appearing in Newton's work in the late seventeenth century. Newton equated the force exerted by a fluid on a body with the change of momentum in the fluid due to the presence of the body.* In our case this force is the drag. Inspection of the drag equation shows that the drag is proportional to the square of the velocity of motion implicit in the dynamic pressure. There is no difference if an airplane moves through air or if the air moves against a model mounted in a wind tunnel. In the wind-tunnel test section—the location where the model is mounted—the airflow must be *uniform*. Such a uniform flow has the same speed in the entire wind-tunnel cross section ahead of the model. Only if this condition is met do we duplicate the relative motion of air to model encountered when the airplane flies in a motionless atmosphere. We used the subscript infinity, ∞, to designate this state of undisturbed air for pressure; it can be applied equally to other flow parameters such as speed, and we will frequently use it.

The drag coefficient of an object depends on its shape and Reynolds number. However, for a group of blunt bodies of great practical interest the drag coefficient often remains unchanged through a wide range of high Reynolds numbers. Such bodies are often called bluff; they differ basically from pointed objects. The small effect of Reynolds number on drag applies, for example, to a disk moving at a right angle to the flow, a sphere, or even an automobile. Therefore, with a fixed size of the object and at constant density of the ambient fluid, the drag of the sphere or the car is proportional to the square of

---

*Permit a slight detour—or ignore it—using the calculus. Newton's law of force (force equals mass, *m*, times acceleration, *a* or *du*/*dt*) can be written by

$$F = ma = m(du/dt) = d(mu)/dt,$$

where *mu*, the product of the constant mass and the velocity, is the momentum. The force is then equal to the change of momentum with time.

the velocity of motion. If we *double* the speed of an airplane, we *quadruple* its drag. The function of drag versus speed thus describes a parabola, as seen in Figure 7.1. This basic law will lead us to appreciate the importance of reducing aerodynamic drag of cars and airplanes in light of the power needed to push them through the air.

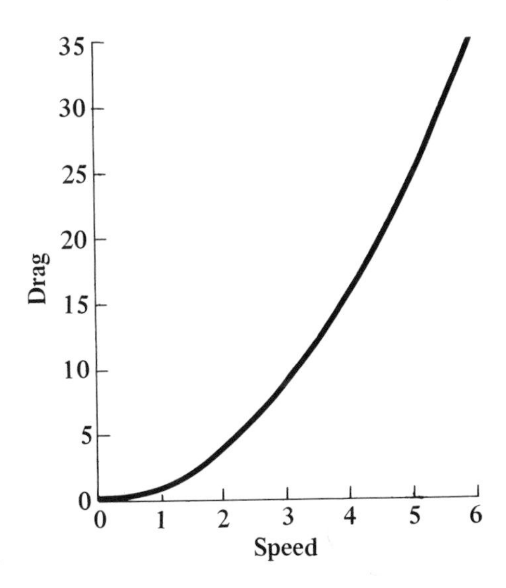

Figure 7.1. Graph illustrating the parabolic drag law (arbitrary scales). The aerodynamic drag varies with the square of the speed.

In general, at Reynolds numbers high enough to get us out of the range of flows wholly governed by viscosity (see below), there are two effects that contribute to the drag of a body. First, the inertial forces cause a variation of pressure such as we discussed for ideal (inviscid) flow, in which we ignored the internal friction (see Chapter 5). The fluid is pushed around by the object in a pattern of streamlines. From the equation for the conservation of volume (or mass) and Bernoulli's equation, together with the geometry of the object, we can compute the flow field. High pressures occur where the fluid hits a blunt surface and slows down, or even comes to a full stop, such as at the nose of an airplane; low pressures appear where it speeds up near surfaces more aligned with the direction of flow. The distribution of pressure about a flying body can be integrated, and it is this summation that leads to the puzzling result that the body shows no drag, as noted in Rayleigh's remark quoted at the beginning of this chapter.

However, the distribution of pressure calculated for ideal flow indirectly dominates the viscous effects; a boundary layer is formed close to the surface (Figure 5.9), and a friction force is exerted that acts tangentially to it. The behavior of the boundary layer is dependent on this pressure distribution, which in turn is affected by whether the boundary layer stays attached to the body. The interplay of inertial and friction forces and their relative importance is controlled by the shape of the object and the Reynolds number. Intuitively we guess that the drag of blunt bodies is dominated by inertial forces, while longish, slender shapes have a higher component of surface friction. Think of the difference of the flow about a ball and a trout. The blunt body will most likely have a flow pattern like that of the flat disk shown in Figure 7.2. Here the flow breaks away from the edge; it cannot possibly move around to the back, where a churning, turbulent motion exists not unlike that behind a rowboat. This breakaway of the fluid is called *flow separation,* the previously mentioned phenomenon that we encounter frequently in the world around us.

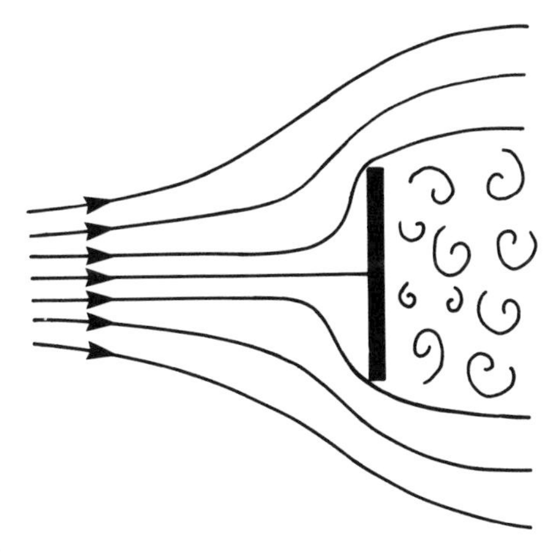

Figure 7.2. Flow about a disk. Separation of the flow occurs at the edge of the disk.

For blunt bodies, the behavior of airflow in the front and the separation in the rear dominate in varying ways. The entire streamline pattern will remain relatively unchanged with speed—that is, with changing Reynolds number. Surface friction will play a minor role. The parabolic drag law applies directly, a condition which is confirmed by experience. We can now look at blunt-body drag without first understanding the friction on surfaces and the details and causes of separation. Typical examples of measured drag coefficients of

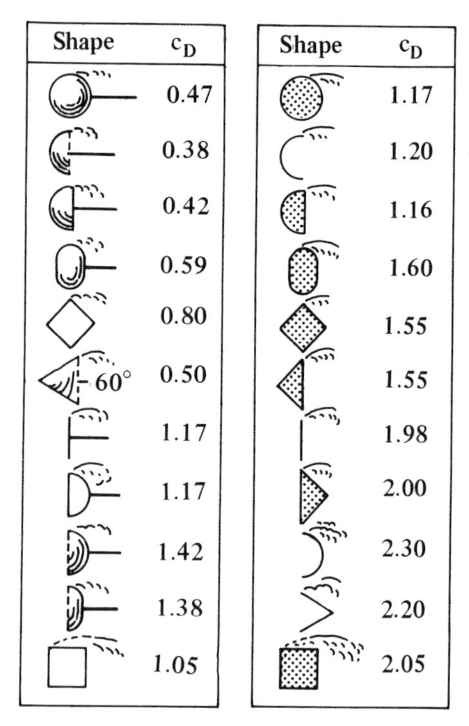

Figure 7.3. Drag coefficients of blunt bodies. Three- and two-dimensional bodies are shown in the left and right columns, respectively. The Reynolds numbers (based on diameter) range from 10^4 to 10^6.

two-and three-dimensional blunt bodies are shown in Figure 7.3. In the left column are three-dimensional objects, such as a sphere and a cube. Take the two hemispheres near the top; the solid hemisphere and the hemispherical shell that is open in the back have drag coefficients of roughly 0.4. Turn these two by 180° so that the opposite side faces the wind, and you will find much higher drag values, as seen near the bottom of the column. For the open hemispheres, numbers two and nine from the top, the drag is nearly four times higher if the "back" faces the airflow. This fact is put to good use in the design of *anemometers,* instruments with which meteorologists measure wind speeds. Four open half-cups mounted in the same orientation on crossbars consistently turn in a single direction, since the drag of an open cup is so much higher. Wind speed can be read from a calibration of the revolutions per unit time versus an absolute measurement taken with a Pitot-static tube (Figure 5.7).

To calculate absolute drag values from the coefficients given, we proceed with the drag equation given above. Aside from the known drag coefficient (e.g., Figure 7.3), we must know the area S. For three-dimensional shapes, we use the cross-sectional area. For two-dimensional bodies such as a cylinder (the top object in the right-hand column of Figure 7.3), we settle on a cross-sectional area for a certain length of the cylinder, like a broomstick held at a right angle to the wind. A cylinder in cross-flow—like an airfoil—has a two-dimensional flow field, since the streamline pattern is identical wherever you slice the flow in the direction of the flow. Here we must be careful. The given drag coefficient works well for the long broomstick, since its value applies to an infinitely extended cylinder. The measurements of the drag are made in a wind tunnel, with the model spanning the test section. However, for short sticks we have *end effects,* complicated flows about the edges that alter the drag. This problem will occupy us when we view finite wings, the real wings of an airplane.

The dynamic pressure, as before, is calculated from flow speed and fluid density. We now have the actual aerodynamic drag force of a flying object and are ready to make engineering calculations. As always, we need to be consistent in our use of units. With the known drag force and other parameters of the flight conditions, we can calculate the power needed to propel an airplane at a given speed, or we can determine—in conjunction with laws of mechanics—the trajectory described by a golf ball or an artillery shell. Finally, when the drag coefficient is constant at high Reynolds numbers, it is a direct measure of the drag of various shapes, provided cross section and wind speed are the same. This simple result applies, for example, to a direct comparison of the aerodynamic qualities of a group of automobiles of roughly the same size.

7.2 The Strange Case of the Sphere

The aerodynamics of spheres has broad applications, since a spherical ball is used in games like golf, soccer, baseball, and tennis. Figure 7.4 shows the drag coefficients of a sphere and a cylinder as a function of Reynolds number based on the diameter as the characteristic length, customarily designated by Re_d. The two curves look alike, except for the absolute values of the drag coefficients. Note that the drag coefficients and the Reynolds numbers are plotted on logarithmic scales (see Appendix 1). In this fashion, a wide range of numerical values can be accommodated on a single scale. We used this means before in Figure 6.3 to show in a single graph a Reynolds-number range covering the settling speed of near-microscopic dust particles and the flight of huge airships, objects that are many orders of magnitude apart. In fact, the drag coefficients of small particles of irregu-

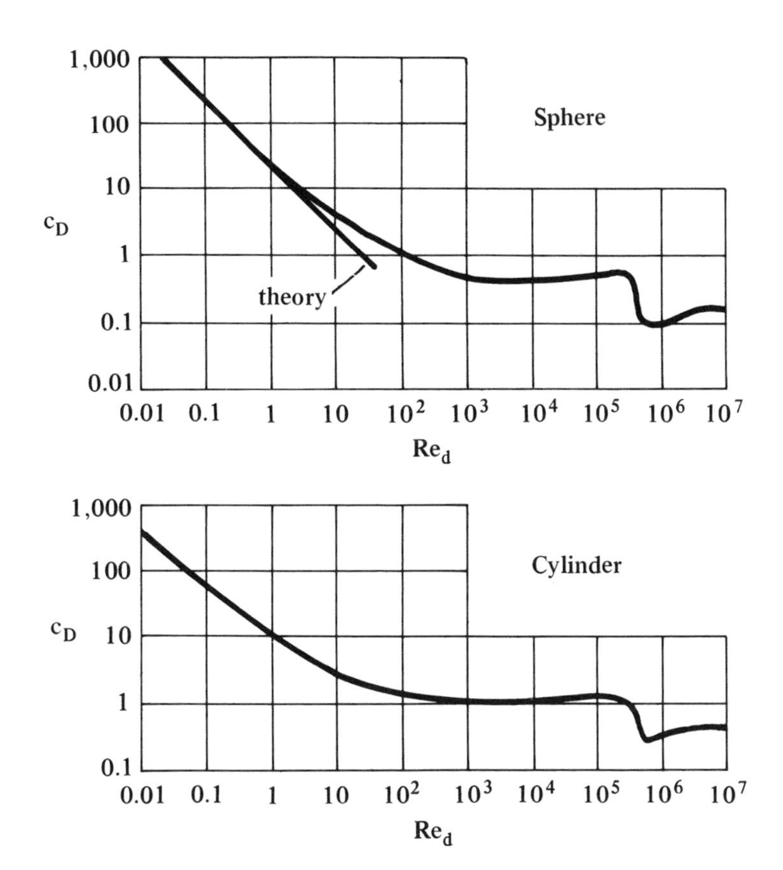

Figure 7.4. Drag coefficients of a sphere and a cylinder as a function of the Reynolds number based on diameter. The curves are derived from experimental results. The straight line for the sphere at low Reynolds numbers is computed for fully viscous (Stokes) flow (see text).

lar shape are well represented by those of the sphere at the applicable low Reynolds numbers to be discussed shortly. The drag curve of a sphere is of particular interest because of the sphere's wide occurrence in man-made objects such as balloons and water towers, not to mention the athletic balls we started off with, and in natural objects such as nearly spherical meteorites. What we will say about the sphere is also qualitatively applicable to the cylinder.

Starting at very low Reynolds numbers around 10^{-2} (or 0.01), we note that the drag coefficient decreases on a straight line in a double logarithmic plot such as Figure 7.4. Here both the x and y scales are logarithms of the quantities shown. (Such a line describes a hyperbola in linear Cartesian coordinates.) Generally speaking, flows at about $Re_d < 1$ (smaller than 1; see Appendix 1), which are called *creeping flows,* are characterized by a low speed of motion, small size of the object, or strong viscous effects. There is no region in the flow at such low Re values where the viscosity can be neglected. No inviscid outer region exists, like in Figure 5.9, and the boundary-layer concept is not applicable, since internal friction of the fluid can nowhere be neglected. Creeping flow about spheres is called Stokes flow after a solution of the viscous equations of motion found in 1850 by the British scientist Sir George Gabriel Stokes (1819–1903). For the drag of a sphere moving with speed U in steady creeping flow with $Re_d < 1$, we have

$$D = 3\pi\mu dU,$$

where d is the diameter of the sphere and π (pi) = 3.14. The drag depends linearly on the viscosity μ: balls fall faster in water than in honey. With the additional linear dependence on the speed U, we note the striking difference between drag at low Reynolds numbers and drag at high Reynolds numbers. In creeping flow we have a linear rather than a quadratic or parabolic drag dependence with respect to flow speed. With the definitions of drag coefficient and Reynolds number we find for creeping Stokes flow

$$c_D = 24/Re_d.$$

This is indeed the equation of a hyperbola, as mentioned before; you can verify it from a high-school math text. (If we take the logarithm of this equation, we arrive at the straight line in the plot of drag coefficients for the sphere in Figure 7.4.) Excellent agreement of experiment and theory exists for the sphere up to Reynolds numbers of about one. The deviation from a straight line at larger Reynolds numbers seen in Figure 7.4 is not surprising; the actual flow departs from fully viscous behavior, and the last equation ceases to be valid. Stokes flow presents one of the rare and beautiful situations in fluid mechanics where an exact solution of the equations of motion is available and, as expected, gives excellent agreement with experiments.

Although we are mostly concerned in technological applications with drag at high Reynolds numbers, the solution of Stokes flow has practical importance. It applies to the motion of tiny particles in liquids, such as the settling of sand in the ocean. Small particles of volcanic ash carried by eruptions into the stratosphere, salt particles carried by storms, dust from disintegrating shooting stars, and even smoke particles from atmospheric pollution all settle to the ground at constant sink speeds given by Stokes's law. Observers of the cloud of volcanic ash ejected into the stratosphere by Mount St. Helens when it erupted in 1980 used this law to estimate the time it would take for the dust to reach the earth. The time estimates permitted them to predict the location of settling according to the prevailing winds in the known circulation pattern of the atmosphere. A final, most important application concerns the behavior of radioactive fallout arising from a possible nuclear catastrophe. The general circulation of the earth (Chapter 4) carries these deadly aerosols; again their sink or settling speed is governed by Stokes's law.

Back to the sphere. Once we have a Reynolds number of about 1,000, the drag coefficient remains constant at $c_D = 0.47$. This is the one case previously alluded to where the drag itself obeys exactly the quadratic speed law (the drag is proportional to the square of the speed) for a wide range of Reynolds numbers, a situation applicable to most blunt shapes. But then, for $10^5 < Re_d < 10^6$, a most remarkable and unexpected effect appears: the drag coefficient *falls* to 0.1! A similar phenomenon is seen for the cylinder. With slightly *increasing* speed, the drag coefficient drops to about one-fifth of its previous value. How do we explain this unexpected and disturbing occurrence? In Figure 7.5a we show the ideal (inviscid) computed flow field about a sphere. With our knowledge of Bernoulli's equation and the equation of continuity, we can draw a curve of the pressure distribution over the sphere. The maximum value of pressure, p_0 at $u = 0$ (see Chapter 5), is reached at the stagnation point for $\phi = 0°$. Strangely, the same pressure value of p_0 prevails in the inviscid flow at the rearward center point $\phi = 180°$. We note the striking symmetry of the curve for the pressure values between 0° and 90° and 90° and 180°, respectively. If we add the pressure forces, we must conclude that no net force acts on the sphere! Since friction is absent by definition in the ideal flow, we find that in inviscid flow a sphere (or cylinder) has no drag! Here is the strange finding that Lord Rayleigh alludes to in the epigraph at the head of this chapter.

This paradoxical result, which clearly violates all experience, was known for a long time, but it took until early in this century to explain what really happens. We know that in fact friction does exist, but from our previous discussion of blunt bodies it does not seem to be a sufficiently large effect to explain the evident air drag on a ball. For example, without drag, a ball dropped from a tower ought to

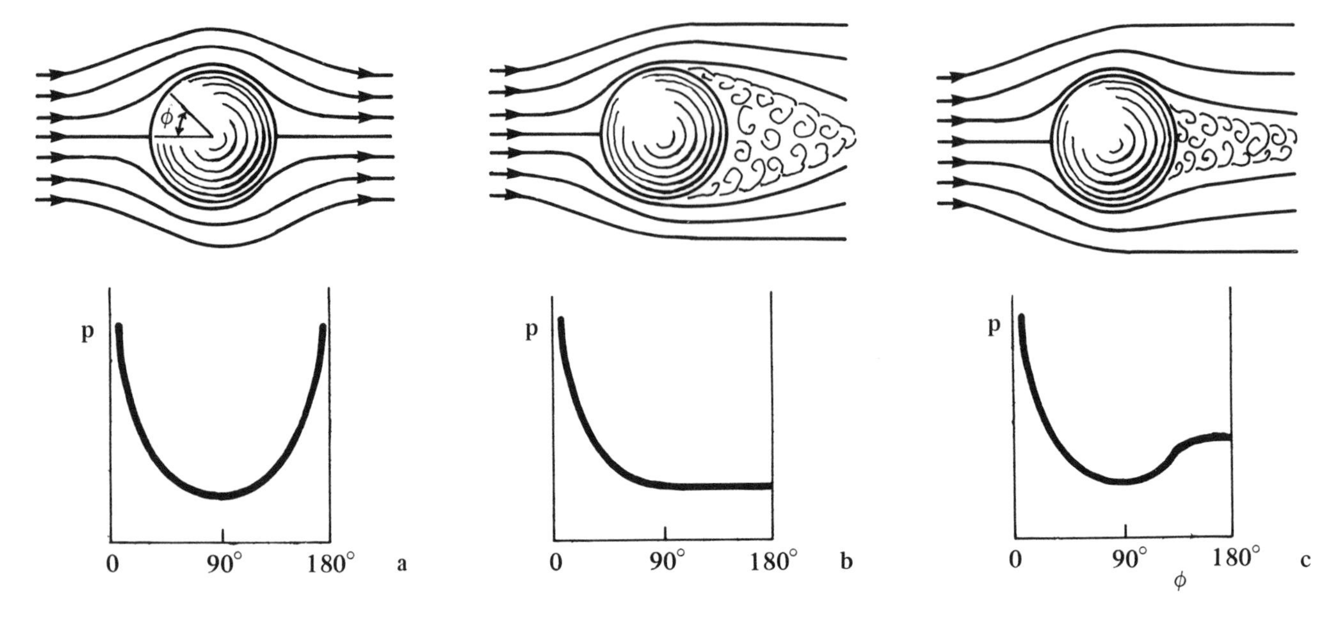

Figure 7.5. Flow patterns and pressure distributions about a sphere. (a) Ideal (inviscid) flow. The angle ϕ(phi) is counted from the stagnation point. (b) Real flow with a laminar boundary layer. (c) Real flow with a turbulent boundary layer.

keep on accelerating as if it were dropped in a vacuum. But we know from observation that it soon reaches a steady speed, just like a free-falling parachutist. Again flow separation provides the explanation of this mystery. While the surface friction on the sphere is indeed very small at the high Reynolds numbers in question, the boundary layer cannot negotiate the adverse—that is, increasing—pressure past the $\phi = 90°$ point. Separation occurs, a wake like that shown for a disk in Figure 7.2 appears, and a skewed or asymmetrical pressure distribution arises, as shown in Figures 7.5b and c. In reality the pressure does not *increase* as predicted for the ideal flow. It remains low in the back, and the frontal forces are not countered and balanced. A net drag force exists to oppose the motion of the sphere.

This leaves us with the final puzzle of the sudden large drop in drag coefficient at Reynolds numbers above about 10^5 seen in Figure 7.4. This mystery was compounded by widely differing test results of sphere drag measured by Prandtl in Göttingen in a small wind tunnel and by Eiffel in Paris, who timed the fall of spheres dropped from his tower. While Eiffel found $c_D \sim 0.5$, Prandtl measured $c_D \sim 0.1$ in the *same* range of Reynolds numbers. By joint efforts the two men solved the puzzle, providing an explanation which is shown in Figure 7.5. If the sphere's boundary layer remains laminar, the separation occurs roughly at the midpoint of the sphere ($\phi = 90°$; Figure 7.5b). However, if the boundary layer is turbulent, the flow is observed to stay attached to the surface to angles larger than 90°, possibly even to $\phi \sim 150°$ (Figure 7.5c). The reason for this difference is that turbulent boundary layers can master adverse pressure gradients—increasing back pressure—more effectively than laminar boundary layers. If the flow remains attached, Bernoulli's theorem takes over, and the pressure increases in the back, pushing the ball forward. Consequently the overall drag is reduced, provided we have a turbulent boundary layer on the sphere. The relatively higher friction drag of the turbulent flow is vastly compensated on blunt objects by the pressure gain, and the drag coefficient drops radically. In fact, the contribution by surface friction to the drag of blunt bodies even for turbulent boundary layers is negligible in relation to the overall drag.

Both Prandtl and Eiffel were right, and both performed good experiments. The smooth air in which Eiffel dropped his spheres delayed boundary-layer transition and the change from laminar to turbulent boundary-layer motion, and thus Eiffel got the high drag associated with the early laminar boundary-layer separation. Prandtl's original wind-tunnel test section had an irregular airflow with high turbulence, and the boundary layer became turbulent at lower Reynolds numbers—it was "tripped." This resulted in the lower drag coefficients. Indeed, transition between the two states (Figure 5.9) can be induced by surface roughness and protuberances as well as by an irregular flow, provided we are in the general Reynolds-number range of incipient transition. Depending on the mode of testing—towing spheres through still air, mounting spheres in wind tunnels with varying airflows or turbulence levels—the drop in drag shown in Figure 7.4 shifts with Reynolds numbers. This important fact invites critical scrutiny of test results in different facilities in the critical speed range in which boundary layers turn turbulent.

We have now solved the strange case of the sphere. If you want to hit a golf ball a country mile, give it dimples!

7.3 Slender Bodies, Skin Friction, Airfoils

So far we have been able to neglect the contribution of surface friction to drag. But this component of the total aerodynamic drag becomes important if we consider long and slender three-dimensional bodies and two-dimensional aircraft components such as wings. Such components are streamlined to produce low values of the drag coefficient, which are essential for the optimum low-drag shape of an airplane or a ship. The lower the drag, the lower the propulsive force needed to push a body through a fluid, everything else being equal. Moreover, smooth flow delays the separation that we found so prominent in the case of the sphere. But let us start at the beginning.

In Figure 7.6 we have two views of a slender body designed for low drag. On the top (a) we note the streamlines of ideal flow. For such flows we build a mathematical model in which Bernoulli's equation for inviscid flow dominates the pressure distribution. The fluid "slips" over the surface, as discussed before in Chapter 5. In real flows, however, we have a boundary layer, as in the bottom view (b), which is "thin" at high Reynolds numbers. In this narrow zone near the body, the flow speed is reduced from the free-stream speed, governed by the inviscid flow, to zero at the surface. The boundary layer, although scarcely noticeable in a photograph of an airfoil flow (Figure 5.3), causes a slight distortion of the actual body shape from the viewpoint of the flow. The object becomes effectively thicker. The boundary layer displaces the flow outward by a value called the *displacement thickness*. This value can be computed from boundary-layer theory, a theory that is beyond the scope of our discussion. The displacement thickness is a fraction of the thickness of the boundary layer shown in Figure 5.9, which was defined as the distance from the surface to the point where the free-stream speed is attained. When studying boundary layers in real flows, we find that the geometry of the streamline picture changes, causing a change in the pressure field, and that surface friction creates a drag effect on the body surface. These two effects must be taken into account for slender shapes like those in Figures 5.3 and 7.6b, since they contribute materially to the total drag. In fact, Figure 7.6b might also be regarded as a side view of the wing section of an airplane.

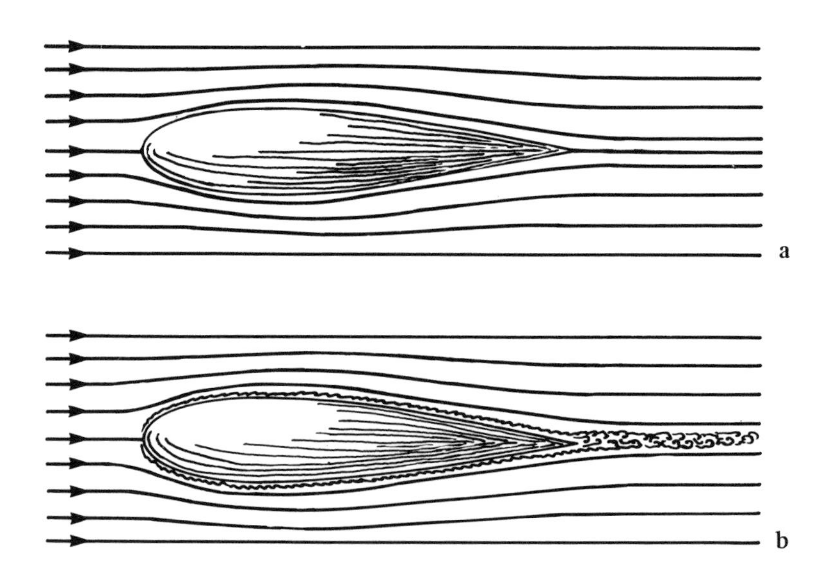

***Figure** 7.6. Flow about a streamlined slender body such as a trout. (a) Ideal (inviscid) flow. (b) Real flow, including a boundary layer.*

We return to the flat plate with the boundary-layer flow shown in Figure 5.9. A laminar boundary layer, a transitional zone in which the flow becomes turbulent, and finally a turbulent boundary layer are apparent. The laminar boundary layer can be treated successfully for flat-plate flow based on the full equations of motion. (Remember that in Bernoulli's equation we neglected viscosity.) The plate is taken to be two-dimensional, and it is further assumed that the pressure outside the layer is constant, an assumption valid for surfaces of moderate curvature. The surface friction can now be computed, and a surface-friction coefficient, c_f, can be defined in exact analogy with the total drag coefficient given before. The area S in these definitions simply becomes the wetted surface area of the flat plate, the area covered by the flow. For the laminar boundary layer the result of these calculations gives

$$c_f = 1.328/(Re_l)^{1/2}$$

for the coefficient of friction. This coefficient is given by the force of friction exerted on a plate of area S, divided by the product of the plate area and the dynamic pressure. We can now calculate the total surface friction of the laminar boundary layer of a plate of area S. The characteristic length, l, in the definition of the Reynolds number is here taken as the distance from the leading edge or tip of the plate. It has been found for engineering applications that the calculation of flat-plate friction drag can be safely applied to slightly curved surfaces of real slender objects. In fact, the calculation is approximately valid for three-dimensional bodies like an aircraft fuselage.

In Figure 5.9 we saw that the transition to turbulent flow occurs at some distance from the leading edge—that is, with increasing Reynolds numbers, here again based on the length from the tip of the plate and designated Re_l. The flow speed and kinematic viscosity used to calculate the Reynolds number are those of the free stream outside the boundary layer. Experimental results tell us that the transition takes place on *smooth* flat plates at Reynolds numbers around 10^5 to 10^6 or higher. The actual Reynolds number of transition depends on the smoothness (or roughness) of the plate, and—as before—in the wind tunnel the turbulence or irregularity of the wind stream enters. Under extremely favorable experimental conditions, transition can be delayed to Reynolds numbers of a million or so. But normally we have the situation shown in Figure 7.7, in which experimental results have been compiled from many sources. The agreement of the theory with experiment is indeed perfect for laminar flow.

Once the boundary layer becomes turbulent, we have only semiempirical treatments of the velocity distribution near the surface and the resulting friction coefficient. This is due to our incomplete understanding of turbulent flows based on first principles. Consequently, the surface friction in turbulent flow must be approximated empirically on the basis of experiments. Most fluid mechanics textbooks provide the sort of experimental data from which the curve of Figure 7.7 for smooth flat-plate turbulent friction coefficients has been copied. Many formulas of varying accuracy are available for surface friction for different ranges of Reynolds numbers. Of these we choose for $Re_l \geq 10^6$ the simple expression

$$c_f = 0.074/(Re_l)^{1/5},$$

which is derived from an empirical curve drawn through the large number of experimental values of friction coefficients on which the mean values of Figure 7.7 are based. Again this expression applies to smooth surfaces. Attaining the required smoothness for minimum surface friction—for example, for an airfoil—is a serious practical problem. Rough surfaces increase the drag and lead to transition at smaller Reynolds numbers. The mix of surface roughness and uneven quality of the wind-tunnel flow is responsible for the wide spread of the experimental results of Reynolds numbers of transition viewed in Figure 7.7.

In comparing laminar and turbulent surface friction, we note two striking facts. In Figure 7.7 we see that at a given Reynolds number where the curves overlap, the turbulent friction leads to much higher

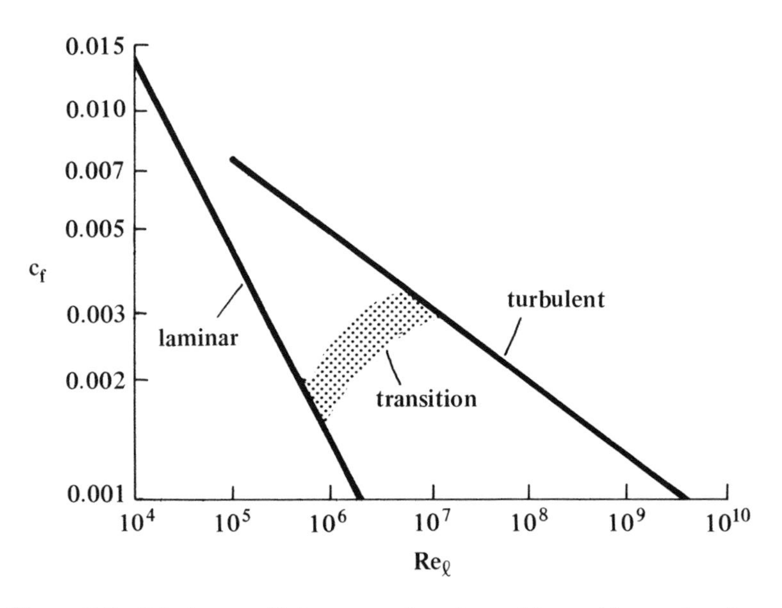

Figure 7.7. Friction coefficients as a function of Reynolds number for a smooth, flat plate computed for laminar and turbulent boundary layers from the equations given in the text. The Reynolds number is based on the distance from the leading edge of the plate. The shaded region between the laminar and turbulent flows indicates the uncertainty of the location of the transition between the two states.

drag values. Next we note that once the boundary layer is turbulent, the friction is much less dependent on the Reynolds number than it is in the laminar situation. We also deduce this fact from the two equations for the two friction coefficients. Mathematically the friction in laminar flow is proportional to $Re_l^{-1/2}$, while $Re_l^{-1/5}$ enters with turbulence. Ideally, we would like to maintain a laminar boundary layer over the entire surface to benefit from reduced friction. This has often been tried, for example on airplane wings, but with little success. There is some hope, based on new ideas, of having laminar boundary layers on airplane wings in the future, as we shall see later, but for the present we will have to live with turbulent flows and their higher friction drag.

To get a feeling for actual values of surface friction and for what its effects on an aircraft might be, let us consider the friction factor on a flat plate for a Reynolds number $Re_l = 5 \times 10^5$. One can achieve either a laminar or a turbulent boundary layer on a plate in a wind tunnel at this Reynolds number. By making sure that little turbulence exists in the free stream and that the plate is smooth, we produce laminar flow. By putting a thin *trip wire* on the plate or roughening it a bit with sandpaper, we get transition and observe turbulent flow. Introducing our selected Reynolds number in the equations, we find values of 0.00188 and 0.00536 for the laminar and turbulent friction coefficients, respectively. Here we assume rightly that the equation given for surface friction is valid for $Re_l = 10^6$. Since the wetted area—the area exposed to the flow—and dynamic pressure are the same for each boundary layer, the plate experiences about three times as much friction drag in turbulent flow as in laminar. If we extrapolate (extend) the laminar curve in Figure 7.7 to higher Reynolds numbers that are realistic for contemporary commercial airliners (see Figure 6.3), the difference of the friction forces on the wing for both flow states will be even more dramatic.

Now let us give some thought to the numerical values of the drag coefficient of streamlined bodies. Shapes such as that shown in Figure 7.6 may indeed have low drag coefficients around $c_D = 0.04$, but for a real object such as a car or an airplane, the idealized streamlined figure cannot be approached as a realistic goal in design. Any actual body of some volume shows values of at least 0.1 to 0.2, be it an automobile (see below) or the fairings of cylindrical struts or the wing-fuselage combination of an airplane. We shall learn more later about the need to compromise the ideal shape. Just open the window of an elegantly engineered and streamlined car, and you'll greatly increase the aerodynamic drag. But now take the extreme case of the high drag coefficient of a disk, $c_D = 1.2$, versus that of the streamlined body with $c_D = 0.04$. At equal Reynolds numbers, these two aerodynamic shapes with equal absolute drag values would have the relative sizes shown in Figure 7.8! This comparison computed from the for-

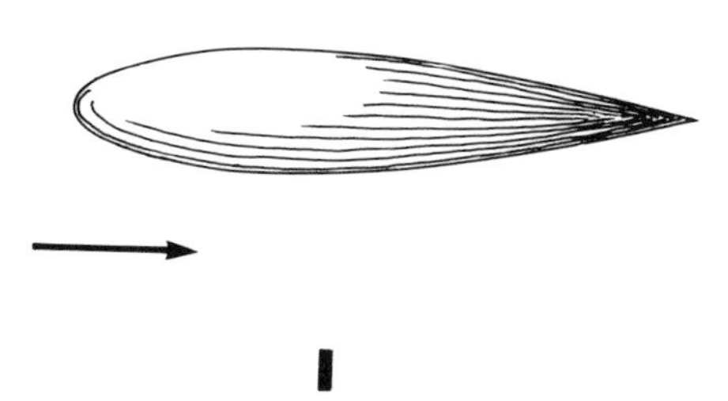

Figure 7.8. *A streamlined, three-dimensional body and a disk that have identical drag!*

mula for drag demonstrates the great effect of aerodynamic shape on the drag of a body, a most important consideration for the design of submarines or airplanes.

The one exception among technically interesting flying machines where separation is relatively unimportant and where friction dominates drag is the airship. Airships have cigar-like shapes resembling those of Figures 7.6 and 7.8. They played a special role in the recent history of flight (see Chapter 2), and some time ago thought was given to a rebirth of these giants as inexpensive freight carriers. However, in view of the success of large transport aircraft, a revival of airships of the size of the ill-fated *Hindenburg* (see Chapter 1) is most unlikely. Their lift—in contrast to that of airplanes—is generated by the upward force of buoyancy based on Archimedes' principle (see Chapter 3.2). Their body is filled with a light gas, usually helium, and their shape can in fact be designed to resemble a huge, low-drag form. Often a fineness ratio (length over diameter) of around 6 to 7 is employed. For airships the unusual situation of predominant friction drag prevails. We shall see later that a modern airliner also has a high frictional-drag fraction of total drag, but the ratio of friction drag to total drag of an airship is not reached. The total drag of the airship body is only about 20% above that calculated for the pure friction drag of the surface of the body. This drag can be computed by our empirical equation of the coefficient of surface friction of the turbulent boundary layer, or we can read the coefficient from Figure 7.7 for the high Reynolds numbers for airships of 10^7 to 10^9. For our purposes it is acceptable to apply friction values for flat plates to this estimate. In other words, the surface area of an airship is turned into a flat plate with a turbulent boundary layer for the purpose of this calculation.

Take the airship *Akron,* whose dimensions are 240-m (785-ft) length and 40-m (132-ft) maximum diameter. Equating the surface of the airship with that of a flat plate, we compute a drag force of 8,000 lbs for a cruising speed of 130 km/h (80 mph) at 3 km (10,000 ft) altitude. To overcome this friction drag alone and keep up a steady cruising speed, a power of 1,700 HP (or 1,300 kW) is required.

Additions to the drag of the body itself arise from stabilizers and rudders, engine mounts and cabin, but they represent small fractions of the total. Here we have the one truly streamlined flying object where flow separation can be delayed to the very end of the airship. It is remarkable that another airship, the *Los Angeles,* with a fineness ratio of 7.23, has the low drag coefficient of 0.071.

We can best understand the details of the effect of *separation* in conjunction with slender-body aerodynamics. If we omit our extreme case of the air-ship and ignore modern airliners for a moment, skin friction normally does not dominate the drag. Yet in a subtle and indirect way the boundary layer does affect the flow field drastically by causing separation. The flow breaks away—it separates—from the body surface; back flow may appear, vortices are formed, and the pressure on back surfaces assumes low values, increasing the drag, as we saw in the case of the sphere and blunt bodies in general.

For a flat plate, we took a constant free-stream pressure outside the boundary layer in the inviscid flow; for a streamlined object, we assumed only a slightly varying pressure. From the continuity equation together with Bernoulli's equation, we know that a constriction in the flow leads to a crowding of the streamlines, thus speeding up the flow and lowering the pressure. Conversely, an increase of area causes the opposite effects. Now every object moving through the air must have a finite length; this is true for a sphere or the streamlined shapes shown. Consequently, the equations of fluid mechanics tell us that past the maximum thickness of an airfoil or a streamlined body the flow must slow down and the free-stream pressure outside the boundary layer must increase. In this region of increasing pressure, kinetic energy is transformed into pressure energy. Boundary layers therefore invariably run into regions of increasing pressure toward the end of a body.

While boundary layers like falling pressure—in fact they decrease in thickness—they have trouble negotiating zones of increasing pressure. The *gradient* of pressure (the change of pressure with distance), in this case its increase with distance, is called *unfavorable* under such circumstances. The boundary layer thickens, the flow at the outer edge slows down, and *separation* may occur. Such a situation is depicted in Figure 7.9. Toward the rear of the object, the surface friction decreases, and the velocity profile near the wall changes. Have another look at Figure 5.8 and the corresponding equations in Chapter 5.3 to note the relation of the velocity profile to wall shear or friction. Take constant viscosity, and note that for a larger angle of

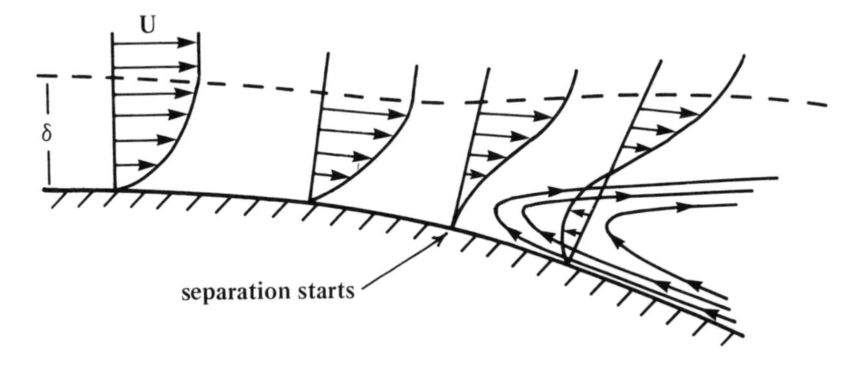

Figure 7.9. *A schematic picture of boundary-layer separation on the rear upper surface of a streamlined object. The increase in pressure caused by the reduction of external flow speed precipitates the separation.*

the velocity profile at the wall, the wall shear—that is, the surface friction—decreases. At some point the flow velocity is zero, even some distance from the surface, and the surface friction goes to zero. At this location the velocity profile at the wall intersects the surface at 90°. Note the point marked in Figure 7.9. This condition defines the point of separation. Past that point the velocity may locally change its direction; the wind blows from the back opposite the free stream, and reverse flow appears. The back flow, shown schematically by streamlines, is accompanied by the formation of vortices, swirling motion, and turbulence. No further increase of pressure following Bernoulli is possible; the flow simply leaves the surface, and the overall drag cannot be further reduced. The aerodynamicist tries to delay this event by intelligent design of the surface shape. Yet practical shapes cannot be long enough to have pointed ends and very shallow angles. *At some stage separation will always appear.* A truly streamlined car would be very long, and it would have a useless tail. A dolphin can do much better; in fact, this animal seems to have a trick to delay separation that we don't yet understand.

The effects of separation are qualitatively applicable to both laminar and turbulent boundary layers. There is one important difference, however. Turbulent boundary layers, owing to more effective momentum exchange by swirling eddies, are more successful in negotiating adverse pressure gradients. They remain attached to the body for a greater distance past the maximum thickness than do laminar layers. Although their surface friction is higher, the sustained attachment permits a higher overall pressure recovery due to the higher pressure predicted by the Bernoulli equation, as we saw for the sphere (Figure 7.5c). This pressure increase may under certain circumstances overcome the higher friction of the turbulent boundary layer even for slender shapes and lead to a *lower* overall drag. Later we shall see how this difference in laminar and turbulent boundary layers affects the retention of lift on airfoils at large angles of attack and other aerodynamic phenomena.

Wherever separation is unavoidable, one may as well chop off the body. This leads to a flat base at the trailing edge of a shape, as we see on objects ranging from projectiles and cars to boats. The so-called *base* drag, which is given by the constant back pressure times that base area, plays a particularly prominent role at supersonic speeds. One can then divide the drag of an axisymmetric body such as an artillery shell into three components: inertial effects—including the drag caused by shock waves—frictional drag, and base drag. But more about this later.

Before we move on to the aerodynamics of automobiles and other earthbound objects, a preliminary note on airfoils. Take a two-dimensional wing in a wind tunnel, stretched from wall to wall just like our infinitely extended cylinder. A symmetrical airfoil that looks in side view just like the streamlined bodies of Figures 7.6 and 7.8 shows its lowest value of drag at the angle of attack $\alpha = 0$. It is simply lined up with the flow. However, well-designed airfoils are not symmetrical; they look more like the wing in Figure 5.3. Their lowest drag occurs at some angle of attack other than zero, a fact that we shall explore when we look at experimental results on airfoils. In each case, a well-streamlined airfoil can have remarkably low drag coefficients of 0.01 or lower, say even 0.005, at Reynolds numbers of the order of magnitude of one million. We note immediately that a treatment of airfoil drag, or the drag of a complete wing, is closely related to its attitude in relation to the three axes of flight of an aircraft shown in Figure 1.3. A car drives on a straight line on the highway, the archer's arrow points its tip into the wind, and the rifle bullet flies straight. Therefore we can give a drag coefficient for a certain design valid for a certain Reynolds-number range. Not so for wings. The drag relates to attitude, and it is intimately connected with lift. It is the lift that sustains the aircraft in flight, and an optimized balance of lift and drag is at the heart of good aerodynamics of airplanes.

Moreover, a real wing is not infinitely extended. It has a finite span, a fact that leads to additional complications of the flow about the wing tip. These problems arise from the pressure difference between the upper and lower surface of the wing, a situation that will soon occupy us. In turn, the lift per unit span is not uniform across the wing from tip to tip. Drag and lift are therefore both affected by the *aspect ratio,* the ratio of the wingspan to the wing chord. In addition to the Reynolds-number range (and the Mach number at higher flight speeds), all these factors—symmetry, angle of attack, and aspect ratio—determine drag and lift. An independent discussion of each force has little meaning; in the next chapter an understanding of their linkage will emerge.

7.4 Automobiles, Etc.

Let us look first at cars. As much as their external appearance has changed since the invention of the horseless carriage in the last century, serious aerodynamics entered design only recently. Why is that? After all, it was long known that drag increases with the square of the speed (Figure 7.1) and, consequently, that cars with low drag coefficients are more economical. But rising fuel costs and the realization of dwindling oil reserves entered our consciousness in the 1970s, a late date considering that automobiles were invented in the last century. Only as recently as that did streamlining, drag coefficient, and other aerodynamic terms enter the automotive vocabulary universally.

There had been sporadic efforts to lower air drag long before the 1970s. In 1921, a German car of aerodynamic design looked like a trout on wheels, including a pointed radiator grille. In fact, the car was called "drop-shaped." Wind-tunnel tests on this full-size car recently carried out at the laboratories of the Volkswagen Company revealed a drag coefficient of 0.28. This is in contrast to the aerodynamically most advanced VW at the time of the tests, which had c_D = 0.34. The 1921 vehicle weighed 3,300 lbs and attained a speed of 105 km/h (65 mph) with an engine of about 35 HP. Another such example in the United States was the Chrysler Airflow of 1934, which followed the trend of streamlining aircraft and trains in the 1920s. It has been noted that the Airflow was an "utter failure in the showroom, but the impact of the design on the shape of the modern automobile has been profound."* Another "aerodynamic" production car, the Lincoln Zephyr, was built by Ford (1936–42); however, except for racing cars, aerodynamic designs did not catch on until manufacturers were collectively nudged by the energy crisis.

The drastic change in appearance necessitated by designing for low drag was initially resisted by the consumer. Now, however, streamlining of automobiles has moved to the fore; the value of the drag coefficient even appears occasionally in advertisements, and contemporary automobiles are beginning to look very much alike. The stylist has finally become subservient to the aerodynamicist. This sameness of appearance is to be expected, since efficient aerodynamic design knows no boundaries between companies or countries. The fact that aerodynamic principles lead to similar appearance has long been known in aircraft development. For example, the British-French supersonic transport Concorde and its Soviet counterpart look like identical twins.

At what stage of driving does low air drag become a sufficient advantage to dominate efficiency? In city traffic, with its frequent starts and stops, aerodynamics plays no role. The force needed to get a car moving from a standstill is given by Newton's law of force: mass of the car times its acceleration. From elementary mechanics (see, e.g., Table A2.2), we recall that the work done in this process is given by the force times the distance covered to arrive at a certain speed. The power expended by the car's engine in turn results from the work done in the time it takes to cover that distance. It is the *engine power* that accounts for the fuel consumption. In a given car you can save fuel by choosing to accelerate moderately, a saving for which you pay a little by lengthening the time interval to attain the desired speed. These simple facts from physics are dramatically visualized by comparing your start once the light turns green with that of a drag racer.

---

*Howard S. Irwin, "The History of the Airflow Car," *Scientific American* 237(2): 98 (1977).

The power needed to reach the supposedly typical city speed limit of 30 mph (50 km/h) depends not only on acceleration but also on the mass of a car. Energy can be saved by building smaller cars and using light materials in their construction. But once the mass is fixed—you have chosen a certain car—acceleration dominates fuel consumption in city traffic. If you want to save money, don't race from traffic light to traffic light.

The resistance to motion of a car at constant speed has three main components: the car's air drag, the rolling friction of the tires on the road surface, and the internal mechanical friction of the engine, wheel bearings, and other parts. The last is by far the smallest of the three contributions, and we shall ignore it. Rolling friction has been much reduced in recent years by improved tire design (the radial tire) and by smoother road surfaces. The aerodynamic drag stands out as the major retarding force opposed to forward motion at constant highway speeds. Its quadratic increase with speed far outruns the moderate increase in rolling friction. Aerodynamics dominates if you drive 40 to 50 mph or so (65 to 80 km/h), and at interstate-highway or autobahn speeds, air drag attains substantial values, suggesting that aerodynamicists must be involved in the styling of an automobile. With everything else (car weight, size, and so forth) being the same, the engine power needed to keep the car going at fixed speed is in fact proportional to the third power of the speed!*

The drag itself is directly proportional to the drag coefficient, provided car speed and cross section are fixed. The range of Reynolds numbers of cars, vans, and buses moving at highway speed is roughly one million to ten million. As we have seen (see Figures 7.3 and 7.4), under these conditions the drag coefficient is nearly independent of the Reynolds number. Consequently, the power needed to keep the car going will be cut in half if the drag coefficient can be reduced to half its original value by advanced aerodynamic design. (Again we must remember that we are speaking of higher car speeds and neglecting the other impediments to forward motion.) All these ways of

---

*Here is a more detailed discussion of this important fact. With drag (D), power (P), and forward speed (U), we have the following relationships. First we recall $D \sim U^2$: the air drag is proportional to the square of the speed. Work equals force times length, or $W = FL$ (see Table A2.2); it is given here by drag times the distance driven. Remember that power is defined by the work done per unit time, here the time to cover the designated distance, and we find $P = D(L/T)$ for the automobile. The speed U is of course given by length over time, L/T, and consequently the power is computed by *drag times speed*. This finally gives the relation that the engine power is proportional to the third power of the constant driving speed—that is, $P = DU$ and $P \sim U^3$, since $D \sim U^2$.

lowering fuel consumption are incentives for manufacturers to produce automobiles of reduced weight and streamlined shapes, and for you to use moderate acceleration and reasonable highway speed.

Let us pause to reflect for a moment on other—often conflicting—requirements of automotive design, lest we overrate aerodynamics. Statistics on accidents tell us that safety improves with the size and weight of the car. The absolute dimensions are also important in many vehicles designed for special purposes, ensuring roadability and stability. This applies, for example, to vehicles that operate on rough terrain, but aerodynamics plays no role in this case. Here, however, we will keep our attention focused on aerodynamics.

Automotive aerodynamics is more complex than one would think. To begin with, it is more difficult to measure aerodynamic parameters of car models than those of airplanes. Airplanes and submarines are fully submerged in air and water, respectively, while automobiles are driven on solid surfaces. Wind-tunnel experiments are performed in various ways to accommodate that difference. A complicated scheme to achieve complete similarity between experiment and reality involves running a car at a fixed location on a movable belt, a sort of treadmill for vehicles. Wind speed and belt speed must be identical, duplicating the actual driving speed. If a full-scale automobile is tested, the Reynolds number is perfectly reproduced by this scheme. Most experiments use models mounted on a fixed plate, but this method falsifies the actual drag values since the model is exposed to the wall boundary layer of the wind tunnel. Empirical corrections of aerodynamic parameters have been established for this and other testing methods. Car models can also be suspended in the free stream, just like airplane models. This of course makes full-scale simulation impractical. There have been other attempts to solve the complicated problem of the relative motions of car, wind, and road, such as having two models glued together symmetrically at their wheels, but we will stop here.

Aside from drag and lift, many other measurements are taken from full-scale automobiles mounted in wind tunnels. With the wind blowing, smoke filaments are directed by probes to various trouble spots where vortex formation originates and separation occurs. Woolen tufts are glued to the body; looking at them we can see if they are aligned with the flow direction or if they jiggle. The distribution of pressure is found by drilling holes all over the car body and attaching pressure gauges on the inside. This method helps designers improve the external shape to reduce the drag. The smoke technique is particularly useful to study the flow around components such as the front grille, side mirrors, antennas, and the like.

A scheme to find the total drag, including the component of rolling friction, under realistic driving conditions on different road surfaces involves the measurement of the *deceleration* of a car. A test driver puts the car in neutral at a preselected speed and steers a straight course. By suitable use of instrumentation, for example bouncing a radar signal off the moving car, a time-distance curve is measured. The deceleration—the decrease of speed with time—can be determined electronically. The total retarding force as a function of speed can then be computed instantly from a solution of the simple equations governing the car's motion.

The aerodynamic drag has a number of components, of which the distribution of pressure about the vehicle—the drag due to shape—accounts for about 85% of the total. Surface friction plays no important role for automobiles. The 85% includes the effects of vortices formed at the rear of the car and other places, as well as interference effects at the edges of the body and around various components that stick out. To this we add an internal drag of 15%, which is made up of the losses incurred by the internal resistance of the air pushed through the cooling systems, ventilation ducts, and the like.

Let us take these components one at a time. The starting point of aerodynamic car design is based on smooth external shapes like those shown in Figure 7.10. From a square plate moving upright along the highway, to a shoe box, to a variety of streamlined shapes, drag coefficients covering one order of magnitude from about 1.2 to 0.12 are observed. One trick to gain low drag is to delay separation; you remember that technique from the discussion of the flow around a sphere. The last three car bodies in Figure 7.10 do just that, in particular the third one from the bottom. Their shapes are similar to those of the symmetrical objects in Figures 7.6 and 7.8. But parking the car with the lowest drag coefficient is not very practical; moreover, no useful interior space exists in the back, and it would not fit in a normal garage.

For realistic automobile shapes we must find a compromise with the unavoidable flow separation. It is useful to fix the flow separation at a given location. We note this in many current models that display an abrupt drop of the roof and a nearly vertical rear window. All the cars shown in Figure 7.10 have low pressures at the top; each car displaces the air, which speeds up, resulting in a pressure drop according to Bernoulli's equation. Observe the bulging top of a convertible, which is pushed outward by the higher internal pressure. Indeed, the car develops aerodynamic lift like an airfoil. At the extreme highway speeds permitted in some countries, the car wants to leave the ground; roadability and control become uncertain. Spoilers—airfoil-like bars—are installed to force the car downward, just as they are now customary on many racing cars. Aside from reducing lift, small wings mounted above the rear window or on the trunk control the back flow in the region of separation. If they are well

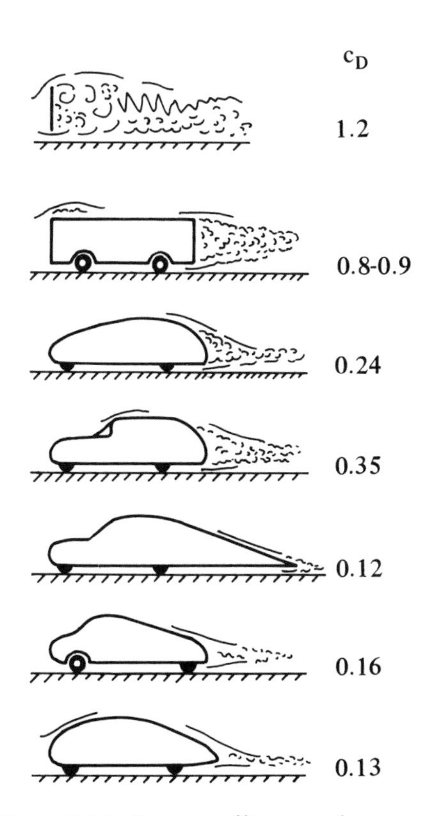

Figure 7.10. *Drag coefficients of smooth automobile bodies.*

designed, these devices can even keep the rear window relatively clear of road dirt.

A real car, of course, is not smooth. It has internal ducting, a rough underside, wheel openings, bumpers, antennas, and other protuberances, all of which add substantially to the drag of the basic, clean shapes of Figure 7.10. Taking all these features into account, we find the average drag coefficient of vehicles on the road in 1985 to be around 0.5, ranging from older models at 0.6 or so down to newer automobiles at roughly 0.4. More recent automobiles achieve drag coefficients of 0.3 or so, and this value may dominate for some time to come. Experimental prototype vehicles, however, have actually reached the much lower value of 0.15. As we have already noted, such low values of the drag coefficient result from radical design changes, which, as we saw in our discussion of the Chrysler Airflow, encounter reluctance of purchasers to participate in the technical revolution. It will be interesting to see which trend will prevail. Oddly, the same reluctance applies to innovations in car safety. Nonetheless, researchers at MIT predict that in the not too distant future the average new car will have a drag coefficient of 0.2 or less.* Be that as it may, it is certain that aerodynamic improvements to lower fuel consumption will be decided primarily by the consumer rather than the technician, since no scientific barriers exist in this situation. Moreover, external political events may again play a role sometime in the future. In view of the fact that a sizable part of the oil used in the United States comes from foreign sources, there may even be a return to automobiles of lower power. With high speed remaining a desirable criterion, aerodynamics will become even more important.

Taking another look at the equation for drag—the actual drag value for a car under given driving conditions—we note that the total aerodynamic resistance to motion at a given speed is determined by the product of drag coefficient *and* frontal area. Frontal area enters therefore as another essential criterion for the designer. The shape of a racing car will not do for us, even if the product of drag coefficient and frontal area is remarkably low. Although a Formula I car has a high drag coefficient of 0.6 or so, it exposes a small frontal area to the wind. The product of drag coefficient and frontal area is only 0.4 m^2, while a medium-sized car with a lower drag coefficient of 0.4 has a total resistance of 0.8 m^2. It is this latter value that counts. Still, a

*The data given are taken in part from Alan Altshuler, Martin Anderson, Daniel Jones, Daniel Roos, and James Womack, in *The Future of the Automobile: The Report of MIT's International Automobile Program* (Cambridge: MIT Press, 1985).

practical vehicle with a drag coefficient of 0.2, rather than the current values, will have 25% better fuel economy at a highway cruising speed of 75 mph (120 km/h) simply because of its better aerodynamics. Such calculations can readily be made with the equations given here and the manufacturer's specifications for a car. Starting with the high drag of the early tall automobiles, which emulated the horsedrawn carriage, much progress has indeed been made.

Closing the discussion of automotive aerodynamics, we note that such work must still play a major role in the future. Aside from drag (and lift), side forces enter into the picture, stability problems arise even at moderate speeds, and parasite drag of sideview mirrors, antennas, and the like needs attention. Taking all these factors together will always be necessary to improve the aerodynamics of a car. Such advances will enhance safety and reduce fuel consumption. Good low-speed wind tunnels are needed for this type of work. In Figure 7.11 we see a ¼-scale model of a car mounted on a turntable in the test section of the wind tunnel pictured in Figure 6.1. The longevity

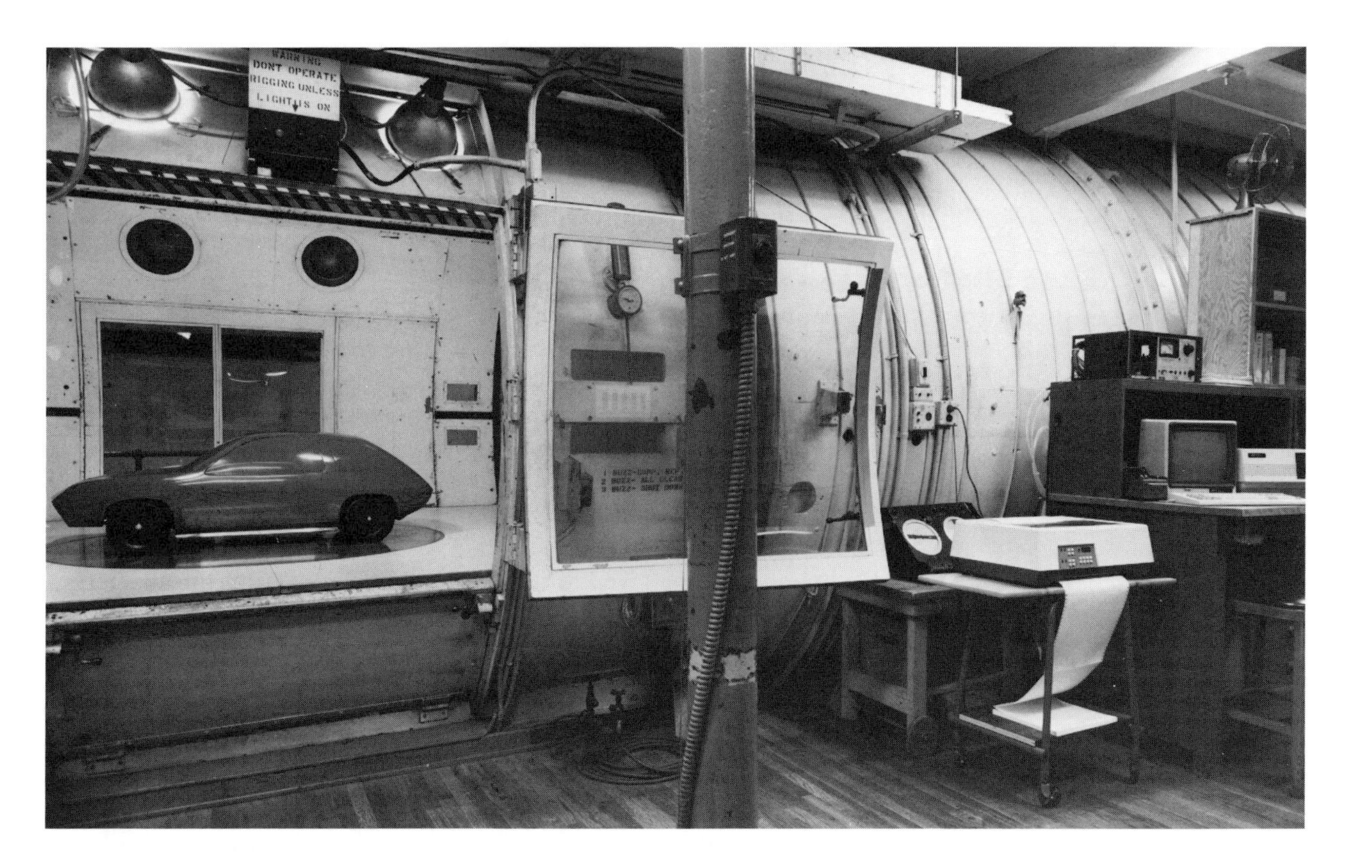

Figure 7.11. Automobile model mounted in the test section of the low-speed wind tunnel shown in Figure 6.1.

of this technical installation is truly heroic; this tunnel has made important aerodynamic contributions, ranging from the Douglas DC-3 passenger airplane and much basic research to the improvement of modern automobiles.

The published data on the drag of trucks and buses are conflicting. Trucks vary greatly in shape, and drag coefficients range from 0.8 to 1.5. The rounded shields on the cabs of tractor-trailers produce a reduction of drag by preventing the separation of the flow at the front of the boxlike trailers. Values from 0.6 to 0.8 are quoted for buses. No serious attempts to streamline trucks and buses have yet been made. A reduction of fuel consumption would seem to be an important goal, considering the high mileage accumulated in long-distance travel by the speeding behemoths that pass us on the highway.

The aerodynamics of railroad trains is important in countries where trains account for a major share of high-speed transportation. As is the case with cars at low speed, air drag does not contribute much to total resistance for regular freight trains. Modern passenger trains are operated by locomotives in streamlined housings that can attain drag coefficients as low as 0.1. The addition of cars obviously increases wind resistance. High-speed trains currently being tested in several countries require aerodynamic research beyond simply lowering drag. For example, aerodynamicists are studying the flow interaction of two passenger trains that encounter each other at high speed in a tunnel. In such situations aerodynamic side forces might derail a train. This work is important in countries like Germany where high-speed trains run at constant elevations maintained by tunnels and viaducts.

Ship aerodynamics has become prominent in recent years. The bulk of the resistance of a ship is caused by the drag of the hull in the water and the energy dissipated in producing the wave system on the surface of the water. Large ships now have bulky protuberances at the bow below the waterline which lower the drag. Model experiments employing additional similarity parameters are produced in water tunnels analogous to our wind tunnels. In addition, much thought is being given to the shape of certain parts of the superstructure above the waterline, in particular the funnels that carry exhaust gases. Wings and fins are attached with the aim of diverting the noxious gases from the decks and ventilation shafts. Funnels are often streamlined, in part for merely aesthetic reasons. In fact, the word "streamlined" has often appeared in descriptions of the design of stationary objects such as toasters; the expression became especially prominent in the 1920s. Much of automotive styling, from hood design to tail fins, followed this trend without any regard for real aerodynamic improvements.

Architectural aerodynamics is another new field. Little streamlining can be done to buildings: the wind can come from any direction, and flow separation at the edges of structures is unavoidable (see Figure 7.2). But an architectural engineer can manipulate the interplay of a number of structures to reduce the draft at street level caused by the atmospheric boundary layer. Other aerodynamic considerations contribute to structural choices such as the exposure of glass surfaces. Models of building complexes are mounted on the walls of low-speed wind tunnels and turned with respect to the wind direction. Fortunately, the actual Reynolds number does not have to be simulated, since the flow will always separate from the sides and roofs of the buildings, and flow conditions are independent of Reynolds number, as in the case of blunt bodies (Figure 7.3).

In contrast, the dynamics of the wind are difficult to study and to predict. They are especially important for bridges and other structures that are in part free from the ground. The interaction of the motion of bridge decks and cables with the shedding of large vortices leads to periodic changes of pressure. If such rhythmic effects are in resonance with the natural frequency of the structure—the frequency with which it would sway by itself if set in motion—the amplitude of the vibration of the bridge can increase to the point of failure. The famous collapse of the Tacoma Narrows Bridge is a case in point. The disaster was finally explained by aerodynamic considerations, which led to the development of improved structures.

But now to close our discussion of aerodynamic drag with a look at the somewhat randomly selected objects listed in Table 7.1. An upright human and some of the other items recall the drag-coefficient values of blunt bodies shown in Figure 7.3. Considering solely the geometric aspects of the different shapes, this ought not to be a surprise. Imagine trying to stand erect facing a wind of 60 mph (27 m/s). A force of about 50 pounds (220 N) will act on you. Most people would be blown over; indeed, sailors cannot stand upright on the fastest destroyers and speed boats. We instinctively lie down flat if exposed to strong winds; by doing this, we reduce the exposed area and the drag coefficient, and we choose the lowest speed in the boundary layer near the surface.

A free-falling parachutist reaches a steady speed of about 100 mph at altitudes from 10,000 to 5,000 ft (3 to 1.5 km). The interplay of gravity, air density, and aerodynamic drag governs this terminal speed. Once the parachute is opened, its higher drag coefficient, primarily resulting from the vastly increased area facing the wind, slows down the fall for a safe landing. The parachute's drag coefficient, like that of falling leaves, depends on the angle of the fall and the parachute's shape and porosity.

TABLE 7.1. APPROXIMATE DRAG COEFFICIENTS OF A VARIETY OF OBJECTS AT HIGH REYNOLDS NUMBERS.

Object	c_D
upright human	1.0–1.3
free-falling parachutist	1.0
recreational skier	1.0
ski jumper in the air	1.2
tall building (Empire State)	1.3–1.5
wires and cables	1.0–1.2
girder bridge, 30% solidity	1.8
Eiffel Tower	2.0
well-designed sea anchor	1.8
parachute (near sea level)	1.4
supersonic rifle projectiles	0.2–0.3
reentry space capsule	1.6
small meteors (outer atmosphere)	4.0–10
pigeon	~0.1
vulture	~0.04

Wires and cables can be regarded as cylinders in cross flow at relatively low Reynolds numbers (see Figure 7.4) owing to their small diameters. We will discuss the drag of objects moving at supersonic speed—that is, at speeds higher than the speed of sound—in Chapter 10, when we leave low-speed aerodynamics. Let us now turn to the more difficult subject of the lift of aircraft, a subject in which drag will again appear as an inseparable part.

Chapter Eight

Aerodynamic Lift

The beauty and vigor of whirling and swirling motions in water and air have fascinated man from the earliest days. Vortices may have inspired Mediterranean artists and craftsmen well over 3,000 years ago to their chains of spiral ornamentations. Vortices obsessed Leonardo da Vinci, who regarded the violent revolving movements as a straight course. Today, vortex motions are still described as the sinews and muscles of fluid motions, and the classical subject of research finds ever new applications: from the structure of turbulent motions to teacup and bathtub whirls; . . . in nuclear reactors and in other energy-conserving schemes; to aeroplanes and other lifting bodies; to wakes behind buildings, moving objects, and oscillating cables; to dust devils, hurricanes, and cyclones in the atmosphere; and to vortex phenomena in cosmic systems.

DIETRICH KÜCHEMANN (1911–76)*

8.1 Early Experience and Some History

We now come to the heart of aerodynamics, the solution to the question of what makes airplanes fly. What is the origin of the force that lifts heavier-than-air machines (Figure 1.1)? To answer this

---

**Journal of Fluid Mechanics* 21: 1 (1965).

question, we shall draw on our knowledge of steady and unsteady flow, ideal (inviscid) and real flow, Bernoulli's theorem, and the boundary-layer concept. Along the way we will discover more about vortices and encounter a lawyer's inspired guess about the nature of lift, arriving finally at a composite picture of what makes flight possible.

The development of the theory of lift has its origins in the last century, and it extends to the 1920s.* Work on the theory runs in parallel with the rapid progress in aviation. It is remarkable that the final touches to round out the physics and mathematics of lift occurred at a time when aircraft were already highly developed and beginning to move toward dominance in long-distance transportation of people as well as the handling of mail and freight. Indeed, the specification of all the fine points of the design of an efficient wing for a particular aircraft is quite recent, a result of the availability of computers and the opening of the new field of *computational fluid mechanics*. The complete solution of the equations of motion about an airplane—including boundary layers—is now within our grasp. As we follow this fascinating history, names of scientists and engineers representing several nations will emerge. Like many other efforts in science and engineering, aerodynamics and aeronautics are truly international.

But let us start by looking at the oldest man-made lifting device. Surely a kite develops lift: you feel its pull on the string. The component of this pull that acts at a right angle to the wind direction, as shown in Figure 1.1, is called the lift. It supports both kite and string. Figure 8.1 shows a simple kite. The flow separates at its edges, and the front of the kite is exposed to pressures close to the stagnation pressure, p_o, on the nose of an airplane. The kite behaves not unlike the disk of Figure 7.2. Kites that produce substantial lift forces—often boxlike structures that give greater stability—can even raise men from the ground. They were used in the last century by armies to provide observation posts similar to tethered balloons.

Kites are not relevant, however, to the design of flying machines. In contrast, early observation of large gliding and soaring birds suggested that slender wings were the answer. Take a series of three simple wing sections, as shown in Figure 8.2. All have the same aspect ratio (the wingspan, measured from wing tip to wing tip, divided by the

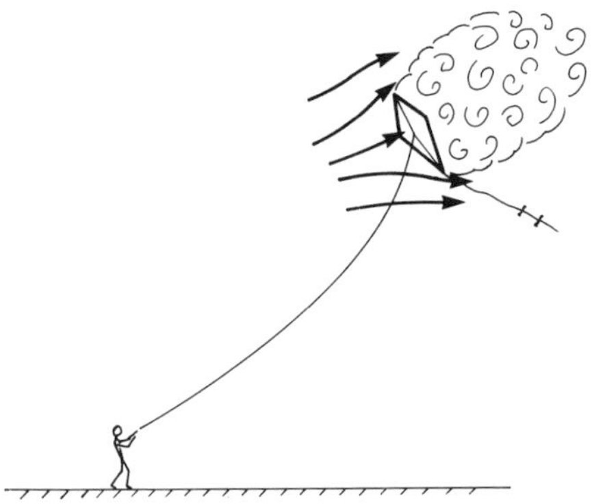

Figure 8.1. *A kite creates sufficient lift at a high angle to the wind to remain aloft.*

*For advanced accounts of the history of the understanding of lift in addition to the books listed in Appendix 4, see, e.g., E.W.E. Rogers, "Aerodynamics—retrospect and prospect," *The Aeronautical Journal* 86:43(1982); I. Tani, "The Wing Section Theory of Kutta and Joukowski," in *Recent Developments in Theoretical and Experimental Fluid Mechanics,* ed. U. Müller, K. G. Roesner, and B. Schmidt (New York: Springer-Verlag, 1979), p. 511.

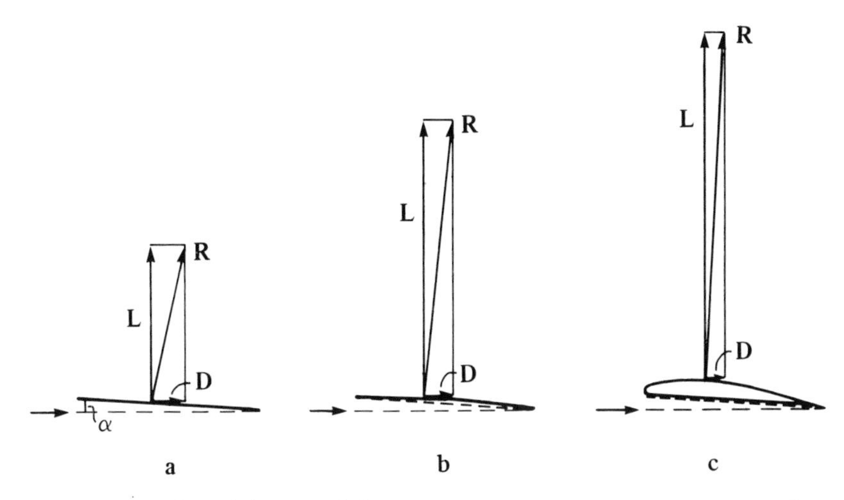

Figure 8.2. Lift and drag on different wing sections. Aspect ratio 6; angle of attack $\alpha = 4°$. (a) Flat plate. (b) Bent plate. (c) Airfoil.

chord [see Figure A2.2]) and an identical angle of attack, yet they demonstrate a remarkable sequence of improved aerodynamic properties. As in Figure 1.1, the resultant air force, the drag, and the lift are indicated by vectors shown as arrows. An increase of lift with decreasing drag is observed moving from section (a)—the flat plate—to section (c). The wing sections shown in Figures 8.2b and 8.2c have *camber* (Figure A2.2), the historical development of which we introduced in Chapter 1, with details and definitions in Chapter 2. The upper wing surface is curved, giving a dramatic increase in lift. Dividing the lift by the drag, we obtain the *lift-over-drag ratio,* a measure of the efficiency of wing design. Lift-over-drag ratios of close to 20 or so are achieved for modern transport planes. The instructive direct physical meaning of this ratio will be discussed in detail later in this chapter because of its importance for gliding flight.

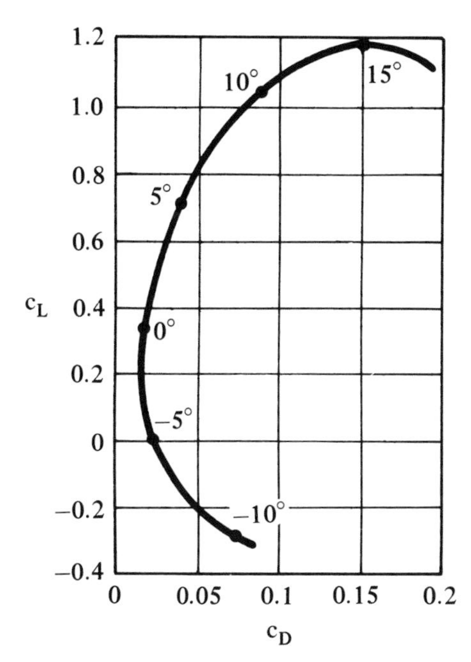

Figure 8.3. A schematic polar diagram of an airfoil. Lift coefficient as a function of the drag coefficient, with the corresponding angle of attack given on the curve (arbitrary scales).

Otto Lilienthal produced plots of lift versus drag coefficients for a given wing at various angles of attack. An example of this type of graph is given in Figure 8.3. Here the lift scale is chosen to be five times as large as the drag scale, reflecting the quantitative disparity of these two coefficient values at a given angle. Such curves, called *polar diagrams,* display the whole story of a given airfoil. A particularly useful aspect of polar diagrams—provided both scales are *identical*—relates to a direct reading of the value of the resultant air force. If we draw a radius vector—a straight line—from the origin of the polar plot (the point where both lift and drag coefficients are zero) to a point on the curve at some angle of attack, this vector directly represents the resultant air force (see Figures 1.1 and 8.2) in direction *and* magnitude. We now have all the aerodynamic information about a wing packed into a single graph!

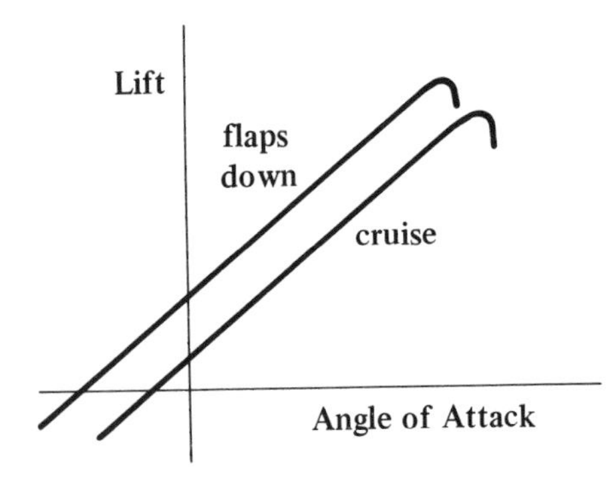

Figure 8.4. A schematic diagram of lift as a function of angle of attack for cruising conditions and with flaps and/or slats in operation.

Lift as a function of angle of attack is shown schematically in Figure 8.4, in which the lower curve represents cruising flight. Most airfoils are not symmetrical about the chord line (Figure A2.2), and consequently the state of zero lift is *not* associated with $\alpha = 0$, an attitude of the craft with respect to the oncoming air where we usually have a positive lift. Increasing the angle of attack increases the lift. This increase is remarkably linear—that is, the lift-versus-angle function is a straight line. But past a certain angle of attack, lift levels off; any additional pointing upward of the airplane's nose—a further increase of the angle of attack—leads to a *stall.* This effect is explained by a separation of the flow on the upper surface of the wing, as shown in Figure 8.5. Although the boundary layer is turbulent prior to separation, a fact that helps to keep the flow attached in the face of increasing pressure (remember our discussion of the sphere in Chapter 7.2), separation at some high angle of attack is unavoidable. Breakaway of the flow from the wing leads to a rapid loss of lift. If the lift force drops below the weight of the airplane, the airplane begins to lose altitude, and the pilot needs to recover an attitude producing sufficient lift.

Again we encounter the two different conditions of attached and separated flow. At low Reynolds numbers we find laminar boundary layers on a wing near its leading edge. As we shall see later, if it were possible to achieve laminar flow on the wing even at the high Reynolds numbers of actual flight, one could take advantage of the much lower friction coefficient to reduce the aerodynamic drag of the whole airplane. On the other hand, as we observed before, the turbulent bound-

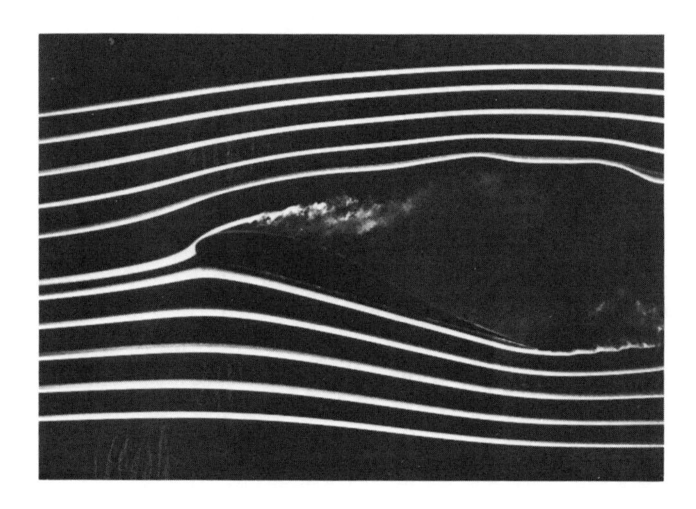

Figure 8.5. Photograph of the flow about a stalled wing (flow separation on top) taken in a smoke tunnel at an angle of attack of 18° and Re = *150,000.*

ary layer on wings of modern airplanes permits a *higher* angle of attack prior to the onset of stall.

To avoid stall, pilots have detailed instructions telling them which flight attitudes are not to be exceeded in a given airplane. Sensors provide warning signals, and skillful piloting makes it possible to fly an airplane out of a stall. But how can we safely provide greater lift for takeoff and landing, for sudden changes of flight path in emergencies, and for other special situations?

In the early days of flight, greater lift was achieved by biplanes, whose two wings are mounted one over the other, as shown in Figure A2.3, and connected by supports, pylons, and wires. Flown by the Wright brothers, biplanes remained popular for many years, making a notable appearance as fighter planes in the First World War. Although the biplane has more lifting surface and higher lift for a given wingspan than the monoplane, or single-wing aircraft, interference effects between the two wings stacked on top of each other increase its drag. At the time it was difficult to build a simple smooth, long wing, sticking far out of the fuselage. The monoplane required a more rigid structure, and it finally became dominant after wood and fabric gave way to metal in the construction of wings. Many different wing combinations and wing-fuselage arrangements are now in use, and some of these are shown in Figure A2.3.

There are other schemes used in practice or tried on experimental airplanes to maximize lift and to increase the angle of attack at which stall sets in. The most common approach uses an extension of the total surface of a wing to increase lift, taking advantage of the fact that lift is associated with the pressure difference between the upper and lower surfaces. This extension can be accomplished by a variety of leading- and trailing-edge devices. During normal cruising flight these movable surfaces are embedded in the wing. They are pushed outward if needed, as shown in Figure 8.6 (see also Figure A2.1). The leading-edge extension is called a slat, or leading-edge flap, and the trailing-edge device is called simply a flap.

On top in Figure 8.6 we see the configuration at cruising conditions. At takeoff (middle), the total wing area is indeed substantially enlarged by a new airfoil shape. To simplify, flaps increase the lift, while slats—depending on their design—increase the stall angle or move the angle of attack of maximum lift to a lower value. This is shown highly schematically by the upper curve in Figure 8.4. In effect, the leading- and trailing-edge devices respectively increase the wing area and alter the wing camber defined in Figure A2.2. An airfoil section is produced that is different from the one most efficient for cruising flight. In fact, that alteration can double the lift for short periods of time during takeoff and landing. More detailed discussions of these subtle effects and descriptions of other devices to increase lift can be found in the aerodynamics books listed in Appendix

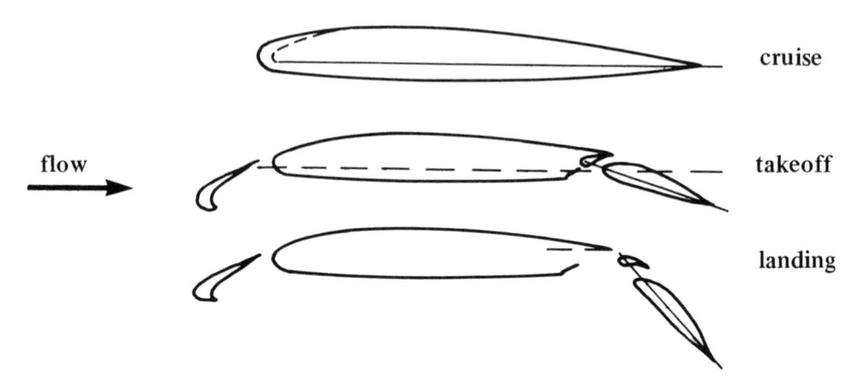

Figure 8.6. Slats (at the leading edge) and flaps (at the trailing edge) added to a wing to control lift and/or stall angle.

4. The underlying principles discussed here are common to all of the devices; yet other principles are involved in unusual methods under study, such as applying suction or blowing on the upper part of the wing and installing movable wing surfaces.

The use of slats and flaps has the favorable side effect of improving control of the boundary layer on a wing's upper surface. The cracks between the movable surfaces and the fixed wing allow higher-speed jets of air to rise from below, where there is higher pressure than above the wing. Because of improved mixing, such jets help the boundary layer stay attached to the wing in the face of increasing pressure toward the trailing edge. This effect can also be induced by suction through a porous upper-wing surface, the experimental method mentioned above that might one day assist short-takeoff-and-landing airplanes (STOL). Such airplanes are being developed to operate at smaller airports that have only short runways.

The bottom sketch in Figure 8.6 shows the landing position of the flaps. They are turned far downward, inducing separation. Exposed below to the stagnation pressure, they retard the speed effectively. Some airplanes also use flat surfaces that are part of the upper wing for a similar effect. These air brakes, or spoilers, can be raised to project at a right angle to the forward motion. Together with a deflection of the jets from the engines to reverse their thrust, air brakes slow a modern airliner on the runway to speeds at which the brakes on the wheels can be safely applied. The undercarriage of the airplane also increases the drag substantially during the landing approach, a fact that makes its retraction important as soon as possible after takeoff. Extension or retraction of the undercarriage, thrust reversal, and the operation of the movable surfaces of the wing produce the various noises that passengers hear during takeoff and landing.

We have noted a few of the many developments concerned with the aerodynamics of wings that do not require an understanding of the actual theory of lift. It is time now to turn to the origin of the lift force.

8.2 Lift of the Infinitely Extended Wing

Looking at the flow about a wing of infinite span is the simplest way to understand how lift comes about. Of course, every airplane has a wing of finite span, which we will have to understand later. But the idealized, two-dimensional flow about the endless airfoil simplifies matters by excluding sideways variations. In practice, just think of a wing spanning a wind tunnel from wall to wall; this is in fact the situation in which airfoils of different cross sections are studied at different angles of attack. The streamlines and the pressure distribution about the wing are identical at any cross section, wherever we slice the wing.

As we saw at the beginning of Chapter 5, a good deal of aerodynamics can be explained by ideal-flow theory, the theory of flows in which we ignore the existence of internal friction, or viscosity. The equation of motion, in this situation the one-dimensional Bernoulli equation extended to motion in two dimensions, applies to the wing.* Computing the streamlines for steady, incompressible flow at speeds much lower than the speed of sound results in the strange pattern shown in Figure 8.7a. There is a rearward stagnation point located on top of the wing near the trailing edge, the mate of the stagnation

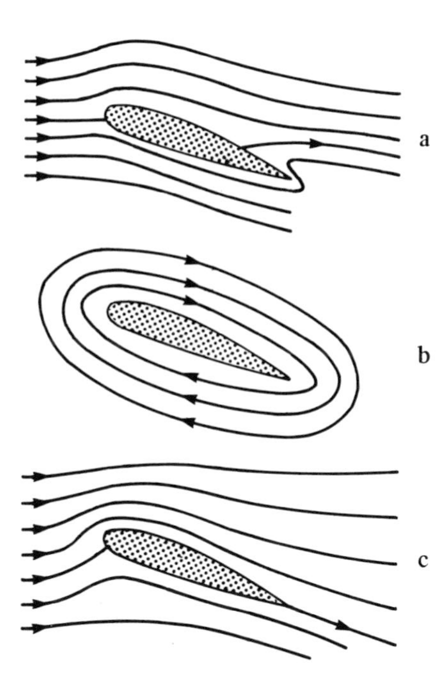

***Figure** 8.7. A sequence of steady, ideal (inviscid) flows about a wing. (a) Calculated basic flow. (b) A circulatory flow around the wing. (c) Addition of the flows shown in (a) and (b), with the strength of the circulation chosen to fulfill the Kutta condition of smooth flow at the trailing edge.*

*Another footnote for the more mathematically inclined. Using the concept of ideal flow, mathematicians have developed an extensive catalogue of streamline patterns of complex situations by the ingenious introduction of a *velocity potential*. The actual flow velocity is expressed as the derivative of this potential with respect to distance in the direction of motion. Using the definition of velocity and applying the law of conservation of mass yields *Laplace's equation,* which is well known in the physics of electromagnetism. This mathematical analogy to flow problems has made available to aerodynamicists an existing arsenal of solutions of Laplace's equation. Choosing proper boundary conditions of a problem, one can compute intricate flows about bodies of arbitrary shape. For example, entire families of geometrically related airfoil shapes can be studied with ease to provide a basis for the selection of efficient wings. In addition to the experiments on airfoils by NACA, the forerunner of NASA discussed in Chapter 2, this technique has been widely applied, giving useful results. All this is possible despite the fact that the airfoils have no drag or lift; as you read on, it will become apparent how far we can go with the ideal-flow concept, as stated by von Kármán in the epigraph of Chapter 5.

point in front where the streamline running into the leading edge has zero flow speed. At both stagnation points we observe the stagnation pressure, p_o, discussed in Chapter 5. Next we determine from Bernoulli's equation the pressure on the surface of the wing from the velocity distribution adjacent to it. Remember that in inviscid flow *there is no boundary layer.* The integration of the pressure around the wing yields the surprising result (mentioned in the note above) that the *wing experiences neither drag nor lift!* The pressure distribution about the wing is balanced to yield no net force. As in the case of the sphere (see Chapter 7.2), this result contradicts all experience: both airplanes and birds fly, producing lift and experiencing drag.

Shortly before the end of the last century, the Englishman Frank Lanchester (1863–1946), a doctor of law who from an early age was fascinated by flight, suggested with uncanny intuition that a vortex acting on the wing is the cause of lift. Lanchester's long life paralleled the explosion of knowledge of aerodynamics. The Aeronautical Society of Great Britain was formed when he was a small child, and powered flight was far in the future. At the time of his death a second world war had ended with the Allied victory based in large part on the advanced state of military aviation. Commercial air transportation had existed for some time, the piston engine and the propeller were being replaced by the jet engine, and the full understanding of lift was providing the basis of the design of efficient airfoil shapes. Lanchester's abiding interest in aviation led him to abandon the law, and his two-volume treatise *Aerial Flight* (1907–08) exerted great influence on the development of aircraft design. In the preface he declared, "The time is near when the study of Aerial Flight will take its place as one of the foremost of the applied sciences, one of which the underlying principles furnish some of the most beautiful and fascinating problems in the whole domain of practical dynamics."

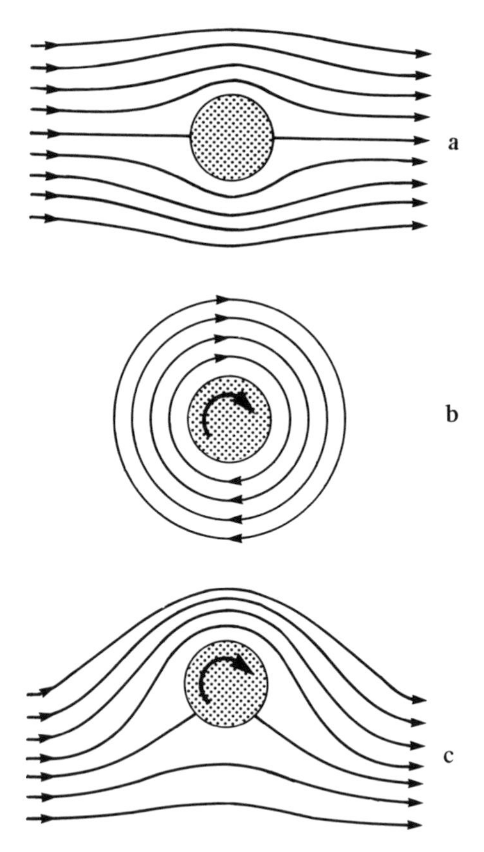

Figure 8.8. *Steady, ideal (inviscid) flow about a cylinder. (a) Flow from left to right. Compare the flow about a sphere in Figure 7.5a. (b) Circulatory flow with concentric streamlines. (c) Addition of the flows shown in (a) and (b). The flow speed is higher on the top, and the pressure is lower; a lift is produced.*

Lanchester's idea of a vortex around the wing causing lift is best understood in light of the long-known Magnus effect, named after its discoverer Heinrich Magnus (1802–70), a German physicist and chemist. Take the ideal flow about a cylinder shown in Figure 8.8a. The symmetrical streamline pattern reminds us of the identical flow about a sphere shown in Figure 7.5a. Once again there is no boundary layer or separation, and the symmetry of the streamlines leads to symmetrical pressures in front and back. In ideal flow, a cylinder in cross flow has no drag. But postulate another ideal flow of concentric streamlines about the cylinder (Figure 8.8b). Superimposing the two flows produces the streamlines shown in Figure 8.8c. Asymmetry now appears in the inviscid flow: the streamlines of flows (a) and (b) move in the same direction on top, and below they oppose each other. Using Bernoulli's theorem together with the equation of conservation of volume in incompressible flow, we find that we have low pressure at the top, high pressure below, and a net lift force acting at a right angle to the flow. By Newton's third law—action equals reaction—a counterforce identical to the lift exists, acting on the support of the cylinder in the wind tunnel, or revealing itself simply as weight if the cylinder moves unsupported through the air.

This remarkable phenomenon of lift on the cylinder caused by the superposition of a circulatory vortex flow and a uniform flow from left to right—the Magnus effect—was verified theoretically by William Strutt Rayleigh, later Lord Rayleigh, the English physicist whose observation about drag we quoted at the beginning of Chapter 7. Rayleigh contributed much to fluid dynamics. He also shared the Nobel Prize in physics in 1904 for the discovery of argon, the gas we encountered as a component of air. The Magnus effect was already well known to artillerists. Just like the projectile of a rifle, an artillery shell is not aerodynamically stable like an arrow, since the resultant air force (Figure 1.1) operates substantially *ahead* of the center of gravity. (We will examine the notions of stable flight in Chapter 9.1.) Firing the shell from a smooth gun barrel causes an erratic and tumbling flight. One purely mechanical method of stabilizing the shell makes use of the stable position one finds in a spinning gyroscope. The similar spin of shells is achieved by rifling in the barrel. Helical indentations act on a soft copper ring about the shell, and the projectile leaves the barrel spinning about its axis, resulting in stable flight. But now the Magnus effect enters: a side force normal to the spin axis and depending on angle and speed causes the projectile to deviate from its planned trajectory. Artillerists ascertain values of this deviation based on range and speed by firing on proving grounds.

Much closer to home, we all know the Magnus effect. A sliced tennis ball has spin; its unpredictable path confuses the opponent. A golf ball too is affected by spin, and last but not least a pitcher's curve ball defies the projection of its trajectory based on gravity and air drag alone.* Let us try an experiment to convince ourselves of the remarkable phenomenon of lift that the Magnus effect produces for a symmetrical body like a cylinder.

Take the light cardboard tube from the center of a roll of paper towels. Make two disks of thin cardboard with a slightly larger diameter than the tube. Glue the disks to the ends of the tube (Figure 8.9). Now wind a string a few times around the center of the tube and place it about one-and-a-half feet from, and parallel to, the edge of a table. (Be sure you have wound the string as shown in Figure 8.9a. The upper surface of the cylinder needs to spin in the direction opposite to your pull for the experiment to duplicate the flow of Figure 8.8c.) Now pull the string vertically downward to move the tube hori-

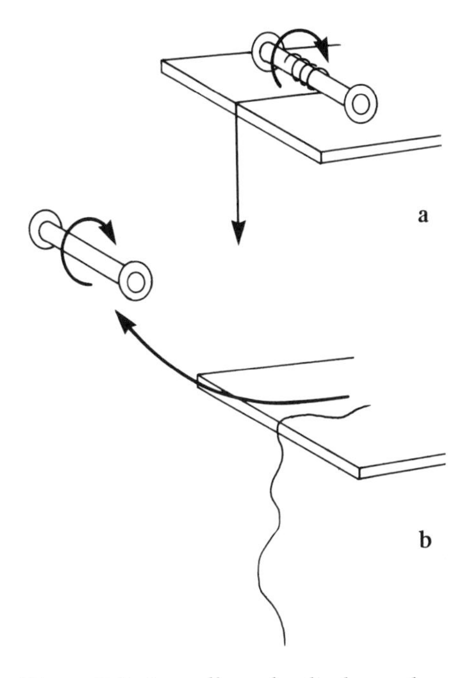

Figure 8.9. A cardboard cylinder and string are used to demonstrate lift caused by the Magnus effect.

*See Robert K. Adair, *The Physics of Baseball* (New York: HarperCollins, 1991).

zontally. The tube begins to spin, moves to the edge of the table, and flies off, arching upward to a height of one to two feet above the table's surface. A little experimentation will teach you to select the proper number of revolutions of string, the right distance from the table's edge, and the strength with which to pull. This simple test is a surprising demonstration of the fact that the combination of forward motion and circulation generates aerodynamic lift that permits a symmetrical object to rise into the air.

In this experiment we did something not strictly in keeping with our notion of lift for ideal flow. By now, I trust, you are convinced that things in aerodynamics are never as straightforward as they appear to be at first glance. The whole theory of lift is an excellent example of the mixture of ideal and real flows, steady and unsteady motion, all combining to explain the remarkably complex circumstances in which heavier-than-air machines can indeed fly. In our experiment the concentric flow about the cylinder—now to be called *circulation*—was set in motion by friction between the surface and the surrounding air. A boundary layer was formed on the spinning tube in a *real* flow. The circulation spread outward, producing the effect postulated for an inviscid flow in Figure 8.8. However, the streamline picture is altered. Although separation is delayed on the upper surface of the cylinder by the action of spin, it does appear in the back, just as shown for the sphere in Figures 7.5b and c. Below the cylinder the ideal symmetry in turn becomes distorted. Nonetheless, the Magnus effect is well demonstrated in our experiment.

An interesting application of the Magnus effect was proposed by the German engineer Flettner in the 1920s. He invented a sailing ship with two vertical cylinders turned by motors acting as sails of a special kind. The direction of rotation and the number of rotations per unit time could be varied, depending on the direction and speed of the wind. The ship could turn on its axis, and in fact, be parked like a car. However, the cylinders had to spin rapidly for the ship to obtain a reasonable speed, and their high drag exerted pressure on the bearings. These problems made Flettner's ship uneconomical.

Lanchester postulated the vortex flow—or circulation—of Figure 8.7b to be added to the cross flow of Figure 8.7a to produce the combined streamline picture about a wing of Figure 8.7c. Looking at the last drawing, and recalling our accumulated knowledge of fluid dynamics, we now understand that a lift is exerted on the airfoil, just as it is exerted on a cylinder. But serious questions remain. What is the nature of the circulation for sustained flight at varying speeds and at varying angles of attack? How does the circulation get started, and how is it maintained at steady cruising speed? Here we again cannot do without mathematics—our equations of motion—albeit expressed in words.

Toward the end of the nineteenth century, the German mathematician Wilhelm M. Kutta (1867–1944) became fascinated by Otto Lilienthal's glider flights. He began to think about the physics of flight, and in 1902 he published his ideas on the subject. This work was carried out parallel to, and independent of, the efforts of Lanchester. Kutta later wrote a doctoral thesis on wing theory (1910). He was followed—again independently—by the Russian scientist Nikolai E. Joukowski (1874–1921), who had built a wind tunnel at the University of Moscow. Joukowski's first paper of 1906, expressing fundamental ideas about lift similar to those expressed by Lanchester and Kutta, was written in Russian. Only a second publication in French, soon to follow, became known farther west. The inevitable sequence of debates and misunderstandings that ensued was fortunately ignored by the practitioners of flight. Airplane designers and aviators had known at least since Cayley's time that wings produce lift, and they experimented with various airfoil shapes to improve it.

Kutta and Joukowski—in contrast to Lanchester, who did not propose mathematical solutions to the interplay of lift and the vortex motion about a wing—put forward the theory of lift in mathematical terms. They quantified the circulation, the vortex surrounding the wing seen in Figure 8.7b.* In simple terms, the *strength* of the circulation can be viewed as an expression of how hard the wind blows on the closed path around the wing. It is that flow which is later to be added to the ideal (inviscid) motion about the wing without circulation. Again note the strange location of the rearward stagnation point on top of the airfoil and the crooked first streamline below the wing that bends upward around the trailing edge (Figure 8.7a). Adding circulation speeds up the flow on top and retards it below. We can push the stagnation point of the ideal flow back to the trailing edge.

In the vicinity of the rearward stagnation point, where the flow comes to a stop, large velocity differences theoretically arise. Extreme flow speeds therefore develop about the trailing edge. It was Kutta who first found a solution to correct this unrealistic mathematical prediction. He proposed a specific choice for the strength of the circulation. In order to have a finite velocity at the trailing edge for given flight conditions, the strength must be chosen to ensure *smooth*

*Their ideas, in mathematical terms, are as follows: The *strength* of the circulation is determined by a nonzero line integral around a closed path surrounding the wing. The integrand (the expression to be integrated) is the component of the local flow speed on the contour—the closed path—in the direction of the aircraft's motion. This velocity component is multiplied by the differential (infinitesimally small) path element along the circulatory streamline.

flow at that location. This situation is shown in Figure 8.7c; the streamlines adjacent to the wing, above and below, have identical speeds when they leave the airfoil. No jump in flow velocity appears. Because it was first enunciated by Kutta, this criterion for smooth flow is now called the *Kutta condition.*

We can now give a final expression for the lift force in steady, ideal flow of the infinitely extended wing. The lift force per unit span of the dimension force/length is obtained from the product of the density of the air, the flight speed, and the value for the strength of the circulation matching the Kutta condition. In the wind tunnel, the flight speed is replaced by the free-stream flow speed in the test section, the speed not yet affected by the presence of a model. The equation is valid, of course, for any fluid, provided the proper density is inserted. For example, the lift of a ship's stabilizers or of a submarine's control surfaces obeys the same laws established here for the lift of a wing. It now remains for us to find the origin of the strange postulated vortex that moves about the wing, and to see how it is sustained. It is remarkable in retrospect that the scientifically untutored Lanchester intuitively suggested the brilliant idea of the circulation about a wing to produce lift. Indeed it is not easy today to follow the sequence of subtle thoughts inherent in the theory of lift. Since these ideas are essential to an understanding of the flight of airplanes, the reader is urged to look at the sources listed in Appendix 4.

The circulation is caused by a special unsteady flow that arises when two streams of different speeds merge at a trailing edge as shown in Figure 8.10a. This figure depicts airflow that was just initiated. Once the flows from both sides leave the object, they move exactly side by side. This cannot go on: different speeds cannot be sustained side by side. The newly formed *surface of discontinuity,* the surface between streams of different speeds, does not remain straight, but undulates in a wavy pattern. Eventually it may roll up to form eddies, as shown in the sequence of Figure 8.10b. The evolution from waves to eddies proceeds from top to bottom. This phenomenon can also occur in the ocean, where a surface of discontinuity may be caused by a jump in temperature or salinity. Similarly, different air masses—cold and warm—if moved about form eddies. As a result we see low-pressure systems, huge vortices that rotate counterclockwise viewed from above and that dominate our storms.

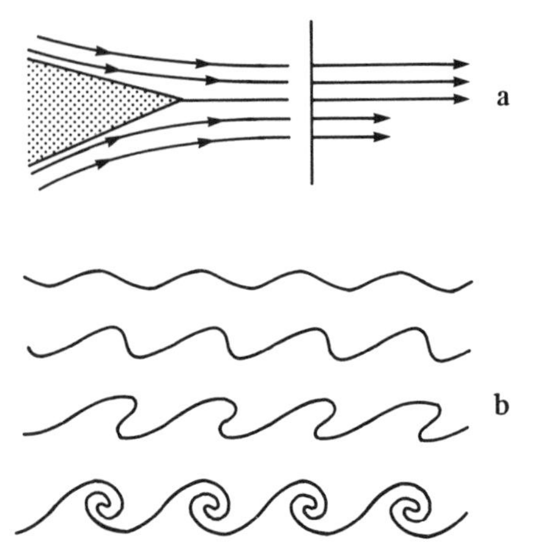

Figure 8.10. *Two flows at different speeds at different sides of an object come together at the trailing edge. (a) A surface of discontinuity is formed. (b) Waves and eddies arising at a surface of discontinuity. View from top to bottom.*

But back to flight. Turning away from the wind-tunnel view of airflow, we will now consider the *unsteady,* time-dependent flow around an airplane wing starting from rest. As the airplane begins to roll down the runway, the ideal flow of Figure 8.7a actually develops. The rearward stagnation point is located on the top surface. It takes some time for friction effects to make themselves felt: the boundary layer grows outward from the surface, and a certain period of time elapses until it achieves its steady state thickness. The air from below the trailing edge of the wing rushes upward to the low-pressure region on top. The difference in velocity between both surfaces leads to very high shear rates, because the flow cannot round the trailing edge at infinite speed. Here the shear—a force per unit area—between the layers of air moving at different speeds, is identical to that discussed in Chapter 5. The resulting surface of discontinuity curls up behind the wing as shown in Figure 8.11a. A special eddy which is called a *starting vortex* is produced, a swirling flow that is quickly left far behind as the speed of the aircraft increases. This second stage of the flow during the starting process is shown schematically in Figure 8.11b. You can verify the existence of the starting vortex by performing a simple experiment. Take a basin of water and pour some ground pepper on the surface to make the flow visible. Now take a spoon and move it sharply sideways on a straight line. The rounded side of the spoon acts like the upper surface of the airfoil, and you will readily detect the starting vortex. This experiment also works with a cup of coffee when you stir in freshly poured cream.

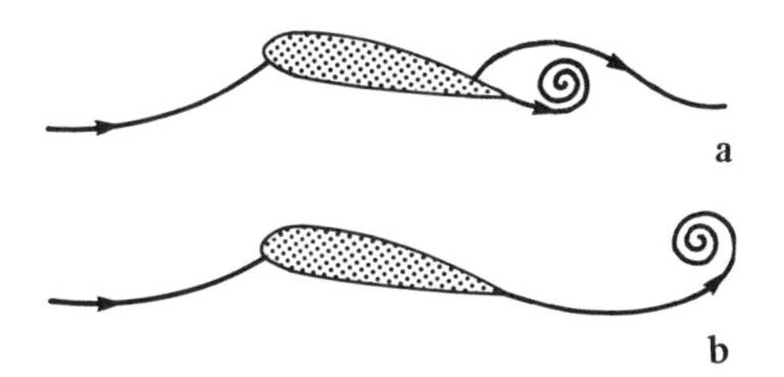

Figure 8.11. *The starting vortex at the trailing edge of a wing set into motion. (a) Formation of the vortex. (b) The vortex is shed and left behind.*

The classic experiments on the origin of circulation were performed by Ludwig Prandtl in a laboratory water channel. Two of his famous photographs are shown in Figure 8.12. Prandtl sprinkled aluminum powder on the surface of the water to visualize the streamlines and trace the motion of the fluid. He produced a uniform stream of water with a paddle wheel. In the top picture, we see the starting vortex in its earliest stage, close to the trailing edge. It is this vortex which soon leaves the airfoil. The physics of the system (another one of those conservation principles!) demands the formation of a second vortex. This additional vortex has a circulation equal and opposite to that of the starting vortex. The specific conservation law that dominates the flow picture is that of the conservation of angular momentum. We encountered the definition of momentum in Chapter 7. There it was linear momentum, the momentum along a straight path as traveled by a car. Here the angular momentum is equal to that of a spinning wheel. But back to the wing. This second vortex works *clockwise* in our framework, and produces the lift first suggested over ninety years ago without an understanding of its complicated origin.

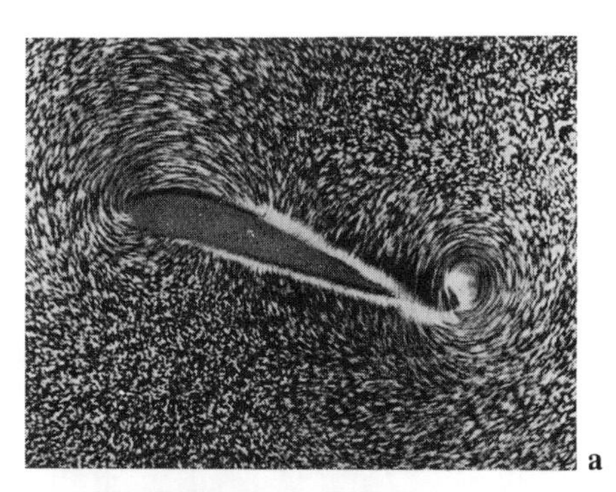

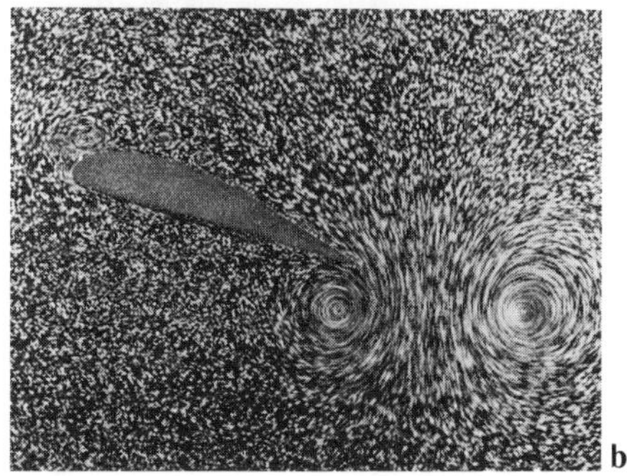

Figure 8.12. *Two photographs of the starting vortex and its mate taken by Prandtl. The flow in the water channel is made visible by aluminum particles sprinkled on the surface. (a) The airfoil is abruptly set in motion. (b) The airfoil is abruptly stopped.*

The mate of the starting vortex—so far postulated by physical intuition—becomes visible when the airfoil is brought abruptly to rest. In Figure 8.12b, we see that the initial eddy has been left behind and its twin has appeared as predicted. The second vortex turns clockwise, like the circulation in Figure 8.7b. To make the second picture, Prandtl started and stopped his airfoil in quick succession. When you fly from New York to San Francisco, on the other hand, the starting vortex is deposited in New York, and its mate appears on the runway in San Francisco. The distance of less than a chord in Figure 8.12b has now been stretched to about 3,000 miles!

The last question to be answered regarding the infinite wing concerns the mechanism that sustains the circulation at constant flight speed in the steady state. To understand the *initiation* of this motion we abandoned the steady state. We return to it now, however, and for the first time in our examination of lift we will look at the real fluid, the *viscous* fluid with internal friction. Figure 8.13 shows a section of the boundary-layer flow on top of the wing. At the high Reynolds numbers of flight of large airplanes (remember Figure 6.3), the boundary layer is turbulent, and it is thin compared to the wing's thickness. Consequently, a significant shear stress prevails adjacent to the surface, where the air sticks to the wing, causing surface friction. The resulting *vorticity*, or prevailing vortical motion, whose direction is indicated in Figure 8.13, induces a continuous downflow that operates in the direction of the circulation about the entire wing and keeps the latter going. The origin of this turning motion becomes clearer if we look at the velocity profile of the turbulent boundary layer shown in the middle of Figure 8.13. The flow speed at the outer edge of the boundary layer is much higher than that near the wing surface, resulting in the swirls (clockwise in our view) that are related to those shown in Figure 8.10. Although there is an opposing motion in the boundary layer of the lower wing surface, it is very much weaker. If the aircraft alters speed or course, the fixed pattern is temporarily changed, but the Kutta condition and the steady state are quickly reestablished by the smoothing action of the boundary layers at the trailing edge on both sides of the wing.

The circulation around the wing is a vortex of a special kind. It works differently from the circulation in rings blown by a smoker or produced by an artillery shell leaving the gun barrel.* Smoke rings

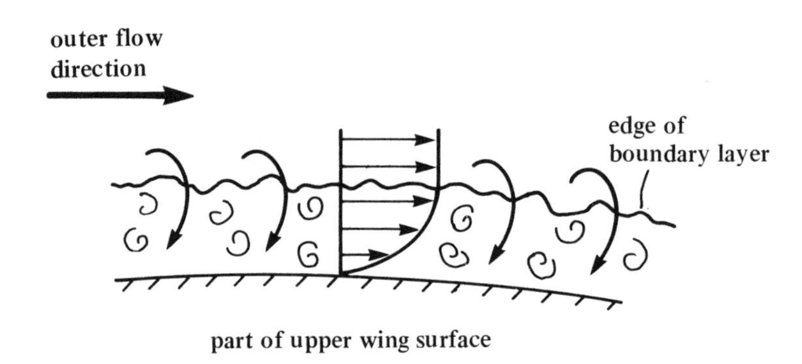

Figure 8.13. Strong shear rates in the turbulent boundary layer on top of a wing create vortices; these eddies feed the circulation around the wing.

---

*A nice experiment to demonstrate smoke rings in the absence of an experienced smoke-ring blower can be rigged up as follows. Put a hole roughly the

float through the air in the shape of a doughnut, curling about the circular centerline from the inside out. This vortex is made up of air marked by smoke, and the fluid elements present at its birth travel with it. The circulation about the wing, which is radically different, is called a *bound vortex,* because it is *bound* to its point of origin. In the steady flow in a wind tunnel it stays at the same place in space, since it is bound to the wing spanning the test section. In flight it moves with the wing of the airplane. In contrast to the smoke rings, the fluid elements—that is, the bits of air making it up—are continually replenished by the oncoming wind.

At last we can compute the distribution of pressure about a wing. With these data in hand, we can determine the resultant air force by integration and find the lift as in Figures 1.1 and 8.2. Applying ideal-flow theory in conjunction with circulation and the Kutta condition leads to remarkable agreement with experiment, as demonstrated in Figure 8.14. Here the results of wind-tunnel tests of a certain airfoil are compared with calculations. The pressure coefficient shown in the figure was defined in Chapter 6; it gives in dimensionless form the difference between local pressure and that of the free stream. Pressure is measured at certain locations on both sides of the wing. The calculated and measured pressure coefficients are plotted versus distance on both wing surfaces from the leading to the trailing edges, x, divided by the chord length, c (see Figure A2.2). The largest excursion of the pressure from the ambient pressure appears on the upper side of the wing. Here the pressure coefficient is negative, since the actual pressure is less than the ambient pressure. Below the wing are positive pressure coefficients. Adding up the forces on both sides of the airfoil yields the net lift to sustain it in the air. At the leading edge we have the stagnation pressure from Bernoulli's equation $c_p = 1$, as demonstrated in Appendix 2 and noted in Chapter 6. The rapid drop to low pressure on top of the wing immediately past the leading edge is remarkable. It indicates a well-designed airfoil. The proper shaping of the upper surface of the wing—the choice of optimal camber—is responsible for the large contribution to total airfoil lift localized near the leading edge. The devices added to the wing shown in Figure 8.6 alter the overall shape of the wing and affect the pressure distribution shown. A minor discrepancy of theory and experiment occurs near the trailing edge. Inviscid theory ignores the boundary layer, but as we saw before, even on a well-streamlined shape separa-

---

diameter of a pencil in the center of the metal, plastic, or vellum head stretched across the shell of a toy drum. Fill the drum with cigarette smoke through the hole. Produce a sharp beat with a drumstick on the other side. A doughnut-shaped smoke ring will emerge from the hole, with the smoke turning from the inside out.

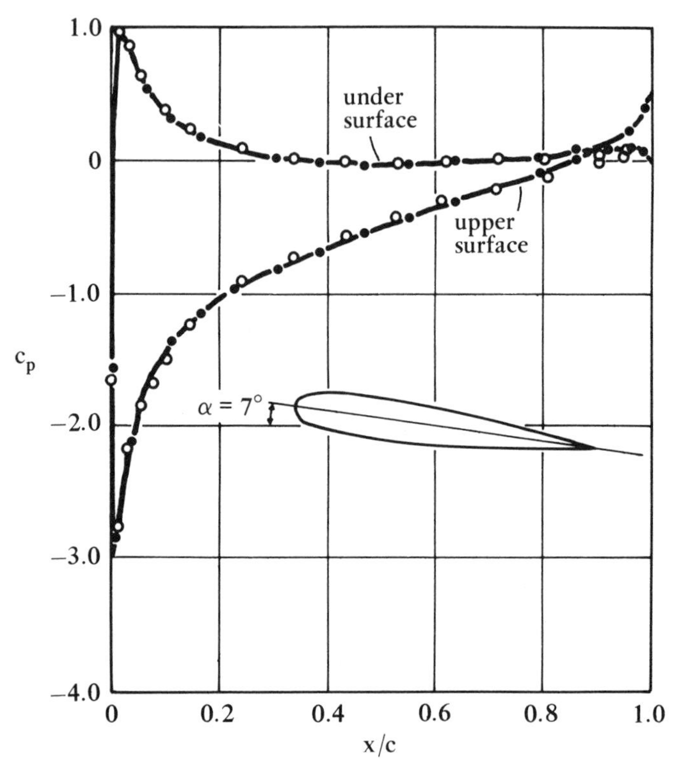

Figure 8.14. Calculated pressure coefficients (dots) and measured pressure coefficients (circles) as a function of dimensionless distance from the leading edge for an airfoil at $\alpha = 7°$. Note the abrupt change of pressure near the leading edge.

tion is at last unavoidable. Nonetheless, the results of the theory permit reliable calculations of airfoil properties. This beautiful theoretical result is responsible, in conjunction with modern computers, for the highly refined wing design of modern aircraft. The advanced airfoil shapes that yield high lift and low drag at the speeds of current jetliners are all based on the same basic ideas. We shall look at such wings in Chapter 10, when we discuss the high flight speeds of modern airliners cruising at about 80% of the speed of sound. At that point, air is no longer an incompressible medium, and additional flow effects will have to be considered.

We have seen that the theoretical basis of the aerodynamics of flight was set out in the remarkably short period from about 1900 to 1904. The theory of lift on infinite wings in inviscid flow was quantified in 1902, and ideal-flow theory was reconciled with observations of real flows by means of the concept of the boundary layer in 1904. In the middle of this period the Wright brothers achieved the first flights in a controlled, heavier-than-air flying machine powered by propellers and a light internal-combustion engine. But questions of whether the vortex around an airfoil really exists, how the finite wing of an airplane behaves, and how drag can be explained were answered over a much longer period of time—into the 1920s—again with Lanchester and with Prandtl and his colleagues in the forefront.

8.3 The Finite Wing of an Airplane

We now cut a finite wing of a certain span (width) out of our infinite airfoil. This real wing has two ends—the *wing tips*—whose effects on the previously described flow must be expected. The difference in pressure between the upper and lower wing produces vortices that are shed from the wing tips as the air from below turns upward. This can be seen in Figure 8.15, a photograph of a crop duster. The spray roughly marks the flow, and two huge vortices hug the wing tips and follow the aircraft far downstream. The combination of starting vortex and wing-tip vortices creates a dangerous downwash—the flow in the wake of an aircraft—that must be avoided by a small airplane following a large one on the runway of an airport.

Figure 8.15. *Wing-tip vortices on a crop duster.*

The energy required to swirl the wing-tip vortices produces a major portion of the drag force of the wing. This *induced drag*—or drag due to lift—is a direct consequence of the flow that produced the lift. The other component of wing drag is surface friction and the minor flow separation at the end of the wing, both caused by the boundary layer. The exact ratio of induced and friction drag to total drag depends on the aspect ratio of the wing and the general design. In our context we cannot discuss the great wealth of accumulated knowledge now available on the details of flow about real wings. For further pursuit of this fascinating subject, reference is made to the literature listed in Appendix 4. But one more remark is important. A more detailed view of the vortex system tells us to expect a permanent downward deflection of the flow about the wing. The lift force, which is perpendicular to the direction of flight for the infinite wing, is now tilted slightly backward. It still acts at a right angle to the effective flow direction seen by the wing, with the latter displaced with respect to level flight. It appears identical to the lift shown in Figure 8.2 if we turn the figure clockwise by a few degrees, except that the drag component now refers to the *induced* drag only.

We have now added two more swirls—the wing-tip vortices—to the bound vortex, producing a horseshoe-shaped swirl. Since, according to the German scientist Hermann Helmholtz (1821–94), no vortex filament can terminate inside an extended fluid, the vortex must be closed. Helmholtz worked in areas of physics, chemistry, and physiology, but his vortex theorems and his extensive work on acoustics have an important bearing on flight (how do you muffle jet noise?). The needed vortex *closure* of the horseshoe is provided by the previ-

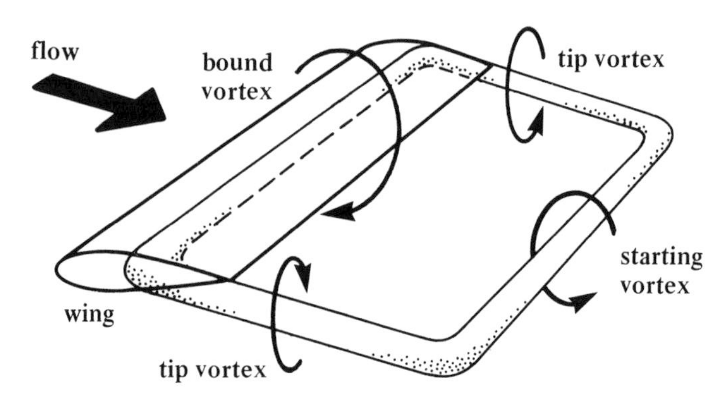

Figure 8.16. The complete closed vortex system of a wing in steady flow. Note that all four vortices turn in unison from the outside inward.

ously shed starting vortex, resulting in the composite picture shown in Figure 8.16. The air swirls in unison from the outside over the top to the inside of the closed vortex. But the system is pulled apart once the aircraft is in flight.

Returning once more to lift, we must deal with the further expansion of the theory to finite wings. We referred at the end of the last section to the ongoing discussions of the subject extending to the 1920s. Prandtl, in another major contribution to the development of the theory of lift, took the entire wing to be a *lifting line* with circulation. He computed the distribution of the lift force across the span and found that the induced drag could be minimized, provided the lift distribution followed an ellipse. The elliptical curve spans the wing, starting with zero lift at both wing tips. Connected with this view are a certain distribution of the circulation across the span and an array of trailing vortices behind the entire wing alluded to previously. Such calculations can be performed to optimize the flow, but an understanding of these matters, essential for the aerodynamicist, again need not concern us here.

8.4 How Birds Fly and Why We Cannot Copy Them

For I think no one believes that swimming or flying can be accomplished in a manner simpler or easier than that instinctively employed by fishes or birds.

Galileo Galilei (1564–1642)

Not only the great physicist Galileo, but Leonardo da Vinci and many others thought that man must emulate the birds to fly. This view changed around 1800, when Cayley, as we have seen, proposed separating propulsion and the lifting surfaces, the wings, and tested his ideas with flying models. Birds, of course, combine the two—their wings provide both propulsion and lift—and the reasons that we cannot copy their mode of flight are now well understood.

What is it that sets us apart anatomically and physiologically from the birds? Even if we had weightless wings attached to our arms, our power output would be much too small to lift us off the ground. The stronger flying birds have flight muscles that amount to about 15 to 25% of their total weight; the breast muscles of the hummingbird raise that portion to 30%.* Our much lower relative muscular weight cannot compare with theirs.

A second major difference concerns the power of bird and man as measured by *metabolism*. This is the process by which energy for the vital activities of the body—including its net power output—is produced. The stored energy is available in the form of fuel such as fat. The fuel is converted by oxidation (breathing supplies the oxygen) into the mechanical work of skeletal and muscular action. This conversion is limited by efficiency factors such as that imposed by the generation of heat; however, for our simple comparison we will consider the net power output only.

Man is best at using energy through his legs. An obvious example is the explosive 10-second power output of the 100-meter dash at the Olympic Games, or the action of a competitive cyclist. First-class athletes can produce approximately 1,000 watts—that is, one kilowatt (kW)—for short periods of time. Think of the power of ten 100-watt light bulbs or that of an average space heater; 1 kW is roughly equal to 1.3 horsepower. During the short duration of the 100-m dash, no fresh supply of oxygen is needed. But a long bicycle race, marathon, or flight in a special human-propelled airplane requires a sustained output of power. Human power output drops precipitously with time, yet a champion cyclist can still produce 300 watts or so for a half hour.

In Chapter 2 we saw that a human-powered, extremely light (but heavier-than-air) craft uses the pilot's legs as an engine to drive a

---

*Data on birds, metabolism, flight, and so forth are taken largely from M. Drela and J. S. Langford, "Human-powered Flight," *Scientific American* 253(5):144 (1985); H. J. Lugt, *Vortex Flow in Nature and Technology* (New York: John Wiley and Sons, 1983), pp. 57–60; J. H. McMasters, American Institute of Aeronautics and Astronautics, Paper AIAA-85-2167 (1984); W. Nachtigall, *Warum die Vögel fliegen* (Hamburg and Zürich: Rasch und Roehring Verlag, 1985); E. R. Nadel and S. R. Bussolari, "The Daedalus Project: Physiological Problems and Solutions," *American Scientist* 76:350 (1988); J. K. Terres, *How Birds Fly* (New York: Harper and Row, 1987); F. Reed Hainsworth, "Energy Regulation in Hummingbirds," *American Scientist* 69:420 (1981).

propeller. Good athletes who can pedal for four to six hours sustain a power of about 3 to 3.5 watts per kilogram mass (W/kg). Assuming a 70% output of their maximum power and a body weight of 145 pounds, such persons produce about 200 to 230 watts for relatively long periods. This was the power output of the pilot who in the summer of 1988 flew the *Daedalus* aircraft 119 km (74 miles) between the small Aegean island of Santorini and Heraklion on the island of Crete. The route of this record-breaking human-powered flight was chosen to repeat, albeit in the reverse direction, the mythical flight of Daedalus and his son, who escaped from Crete. Remember that Icarus, a rash young fellow, flew too close to the sun; his waxen wings disintegrated, and he became the first victim of an aviation accident.

Now compare the relative power of birds. If it comes to a quantitative description of maximum speed, endurance, and similar characteristics, there exists some uncertainty. This is understandable in view of the difficulty of measuring the metabolism of birds flying tethered in a wind tunnel or estimating power output from known food intake and other variables. All details aside, the ruby-throated hummingbird stands out, holding the world record for the highest power per unit mass achieved by a warm-blooded animal. Its mass is about five grams, and in hovering flight—flight at zero speed just like a helicopter—the bird produces and sustains a power output of 16 W/kg. Hovering requires high power; at a flight speed of 14 km/h (9 mph), the power requirement is reduced to 12 W/kg, while at speeds of 50 km/h (30 mph), this value increases to 32 W/kg or so. These high power outputs can be sustained for long periods of time; compare these values to those generated by our cyclists and note that the hummingbird does about one order of magnitude (or ten times) better! This tiny bird, which weighs about as much as a copper penny, can produce ten times the power per unit mass that a man can. You can take a guess at the human power output that would duplicate this remarkable bird's feat.

The hummingbird may be an extreme case even among birds, but it is obvious that human power is insufficient to carry us into the air; evolution has not turned us into airplanes. However, with the extremely lightweight human-powered aircraft, low-speed, level flight has not only become a reality; these flying machines have been kept in the air for hours. Again birds beat this achievement, with migrating birds mastering remarkably long flights: our little hummingbird is known to fly 2,400 km (1,500 miles) without pause. Birds have to feed for an extended time before taking off to accumulate extra fat—that is, to fill their tank with fuel—in preparation for a long-distance flight. For example, a nightingale requires one gram of fat—its original weight is about five grams—to cross the Sahara Desert!

Human aviators, including Lilienthal and the Wrights, have learned a great deal from birds by observing the passive flight modes of *gliding* and *soaring*. Similar conditions apply to airplanes in power-off gliding flight. Besides gliding, modern sailplanes can soar—that is, ascend above their starting level—by taking advantage of vertically rising air masses. Such upward winds occur if air is deflected by a mountain range or rises in convective currents. The latter appear above areas where the air is heated from below; hot air rises, and if the vertical wind speed exceeds the sink speed of the sailplane, the craft soars. Gliding and soaring of sailplanes—aided not only by advanced aerodynamics, but also by lightweight materials—can in fact compete with, or even exceed, the "power-off" flight of birds.

How does gliding work? The aerodynamics of a glide may readily be understood by applying elementary trigonometry. Recall Figure 1.1, where we saw a force diagram of an airplane in powered level flight, with the forces acting in the horizontal flight direction (thrust and drag) and the vertical axis (lift and weight) canceling each other. We noted that if a net force existed, the cruising aircraft would depart from its steady course, slow down or speed up, rise or descend, or bank. Similarly in gliding flight, as shown in Figure 8.17, no net force acts on a bird, or for that matter on an airplane. The coordinates of level flight are now tilted by the glide angle ß (beta). (Birds like glide angles of 3° to 5°.) The bird has the velocity V along the flight path, and a sink-speed component V_v is directed vertically downward. The thrust T is induced by gravity. It is given simply by $T = W \sin ß$, where W is the weight of the bird. The thrust equals the drag; consequently, the sum of the forces along the direction of the flight path is zero, just like in level flight. As required for the steady state—the gliding flight where nothing changes with time, just like cruising level flight—the sum of the forces at right

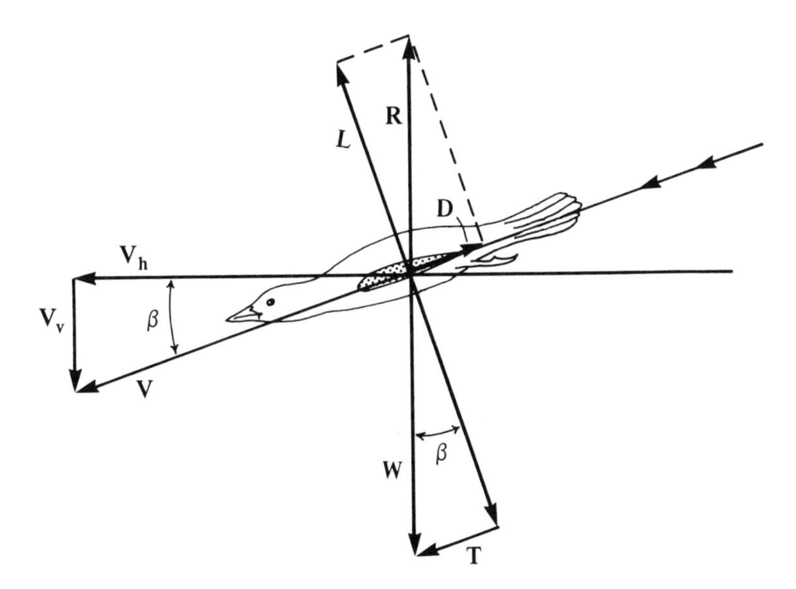

Figure 8.17. Force diagram of a gliding bird (or airplane).

angles to the flight path is again zero, with $L = W \cos \beta$ counteracting the component of weight downward at a right angle to the flight path. The lift-over-drag ratio, which dominates the nature of the glide, is equal to the ratio of the horizontal component of the flight speed V_h to the sink speed V_v. From the previous expressions, remembering $D = T$, we derive $L/D = \cos \beta/\sin \beta = \cot \beta = V_h/V_v$.

Clearly a large value of L/D implies that the horizontal speed, the forward speed parallel to the ground, is much larger than the downward speed. This fact gives the bird (or aircraft) a correspondingly large range before it must touch the ground. Gliding is practiced not only by birds, but also by certain fruits and seeds, such as maple seeds. For airplanes—including sailplanes and hang gliders—the all-important L/D ratio depends on the angle of attack, the aspect ratio, and the specific characteristics of the wing-section profile. Values of L/D (also applicable to gliding) can be read from polar plots of a given wing (Figure 8.3). The inverse value D/L is called the *glide number.* Say the glide number is 1:20 (i.e., $L/D = 20$): then the flying object drops 1 meter while it covers a horizontal distance of 20 meters.

The wandering albatross (*Diomedea exulans*), one of the largest birds in the world, is a master of gliding and soaring (chickens don't soar). With a wingspan of 3.5 m (11.5 ft) and an aspect ratio of over 15, it develops L/D values up to 20. It can fly at speeds up to about 110 km/h (70 mph). Some birds have reportedly been observed to fly up to 100 mph (160 km/h) for short distances. In contrast to the record holders, most birds fly at speeds in the 10- to 60-mph range, a fact that is apparent to a driver who keeps up with a flying bird. When it comes to soaring, the best modern sailplanes, with L/D ratios up to about 60—or glide numbers of 1:60—beat the albatross and similar birds. Hang gliders, with L/D between 4 and 9, emulate pigeons, which operate in the middle of this range. Power-off glides of small aircraft have glide numbers of 1:10 or so. Large airliners cover close to twice the distance of small planes for the same drop in altitude. And the return from orbit to the earth's surface of the biggest sailplane of them all, the space shuttle, is one enormous high-precision glide path in the earth's atmosphere.

The comparison of bird and plane as gliders teaches us that the theory of lift developed in the first three sections of this chapter applies equally to both. The aerodynamic handling of flight by birds, however, displays an unparalleled subtlety. The control about the three axes of motion, from which the Wright brothers learned, is carried out by minute changes in the attitude of the wing tips, the feathers above and below the wing, and the tail. In fact, birds can alter the whole polar curve by literally changing the shape of the wing (acting as an airfoil of variable camber) to adapt it to a required maneuver. It is this adaptation that enables a bird to operate its wings in a fashion that simultaneously produces lift *and* thrust, transforming the entire animal from a gliding to a powered airplane.

The details of the rapid motions of flight are best viewed by freezing an image, as shown in the photographs in Figure 8.18 of a bird in level flight. The force diagram of Figure 1.1 applies; thrust and lift are both provided by the flapping wings. How their complex motions can perform this dual function is much more difficult to explain than the flight of an airplane. In drastically simplified terms, the inner half of the wings—the "arm"—acts primarily as a lifting airfoil. The outer part beyond the wrist joint (the bend visible in Figure 8.18)—the "hand"—provides the thrust. The outer part thus performs the func-

Figure 8.18. *A gull (*Larus ridibundus*) in level flight. The upper and lower photographs show the instant of the completion of the downward and upward beats, respectively.*

tion of the propeller of an airplane (see Chapter 9.2). The angle of attack of the lifting surface is ordinarily about 3° to 5°; at 25° or so, stall sets in. The stalled wing position is assumed for landing, and gaps between the feathers appear, much like the flaps and slats that we discussed earlier. In fact, extra stall feathers act like spoilers when they are raised prior to touchdown.

But back to level flight. The outer portion of the wing, with its rounded leading edge, moves during flapping like a propeller. From our preceding remarks on the lift of airfoils, it is obvious that a wing moved vertically through the air produces lift in the horizontal—or forward—direction. The up-and-down motion of the wings in flight is associated with a highly complex, time-dependent change of their shape. The angle of the inner and outer wing parts changes smoothly; the profile of the wing in each stroke is varied to provide optimum thrust, here lift in the forward direction. During this action the inner part steadily maintains its lift to keep the bird in level flight. The body of the bird is shaped to be "streamlined"—that is, to achieve minimum drag.

To repeat the sequence of events in the simplest terms possible (nothing is really simple about it!), the downward beat of the wings is associated with forward motion of the leading edge, while the upstroke moves the wings backward. Thus the plane in which the wings beat is not at right angles to the direction of flight but rather tilted obliquely backward. The frequency of the beats varies, depending on the speed of the bird and its flight pattern—start, level flight, and so forth. Many birds flutter in level flight in a frequency range of 2 to 5 Hz*; a pigeon normally flaps its wings three times per second (3 Hz). Again, there are wide excursions from these values: the ruby-throated hummingbird, as always, deviates from the pack, beating its wings at 50 to 70 Hz.

The outer wing—the propeller—consistently produces lift in both strokes in addition to the forward pull. During the downbeat the primary feathers, or flight feathers, near the wing tips stand out at nearly a right angle to the flight direction at the point of the highest relative motion through the air. On the upward beat the outer leading edge provides the propeller action, an action that is adjusted to the proper orientation between the two strokes. The wings occasionally rise sufficiently high to touch each other. The secret of bird flight lies

*The SI unit hertz (cycles per second) is named after Heinrich Hertz (1857–94), who provided experimental proof of the existence of electromagnetic waves. Hertz was the star student of Helmholtz, who contributed much to aerodynamics, as we saw in conjunction with vortices.

in the animal's ability to adapt speed, shape—largely camber—and the angle of different parts of the wing to provide an optimum mix of lift and thrust.

A further fascinating aspect of bird flight is the ability to hover in one location—even, in the case of the hummingbird, to move backward. The two wings act again like a propeller or the blades of a helicopter, moving air by "suction" from above to below. Moreover, they create a constriction that speeds up the air by the conservation of mass. The resulting change of momentum (see the beginning of Chapter 7) leads to a vertical upward force, an aerodynamic lift that counteracts weight; the result is that the bird stays in place. The hummingbird produces this lift by downward *and* upward strokes of its wings. The angle of attack of the whirring wings is switched to the opposite direction at the top and bottom of the strokes, and an oscillating drag caused by the forward and backward motions of the wings appears. The forces in the two directions balance each other, and the bird remains stationary. There is a price, however, as the compensated drag forces consume energy without producing lift. Remember that more power is needed to hover than to maintain moderate forward speeds. We can now see that nature has provided a close equivalent to the propeller, in effect a continuously turning wing. It is not quite the same though: the device of a turning shaft has not arisen by evolution in biological systems—that is, nature has never duplicated the wheel.

There are many complicating facets to our simplified picture of bird flight. Wing-tip vortices move up and down, a downwash caused by air deflection exists as for airplanes, and the tail feathers are moved to contribute to the beautiful aerodynamic motion displayed by most common birds. Many of the fine points of their flight are studied intensively by biologists because of their intrinsic interest. These researchers have to be aerodynamicists as well. A better understanding of the periodic distribution of lift and thrust resulting from the motions of a bird's wings and of the details of the bird's flight control also has practical applications for aviation. The way that the boundary layer on the wing becomes sufficiently stirred to remain turbulent and thus to delay a stall at the relatively low Reynolds numbers of birds (see Figure 6.3) is by itself of sufficient importance to keep such research going.

Interesting additional facts of aerodynamics arise in the study of the flight of migrating birds, which often travel in the well-known V-formation. The *upwash*—upward air motion (see Figure 8.15)—of the wing-tip vortices of the lead bird tilts the resultant air force of each follower forward and reduces its drag. A spacing of ¼ wing span optimizes this effect. Theoretically twenty-five birds could save enough power by flying in formation to experience an enormous range increase (possibly 70%) compared to a single bird flying by itself.* The opposite effect arises if an airplane follows another directly behind—for example, when refueling in flight—rather than spaced sideways as in the V-formation: the downwash of the first plane increases the second plane's drag.

*P. B. S. Lissaman and C. A. Shollenberger, "Formation Flight of Birds," *Science* 168:1003 (1970), and J. J. L. Higdon and S. Corrsin, "Induced Drag of a Bird Flock," *American Naturalist* 112:727 (1978).

Chapter Nine

Notes on the Whole Airplane

You, to whom it may concern when I am gone, may find the seeds of thought in these scrawls.

[A NOTE INSCRIBED BY SIR GEORGE CAYLEY (1773–1857) ON THE FRONT BOARD COVER OF AN EARLY NOTEBOOK.]

9.1 Stability and Control

The topics of this chapter are much intertwined with questions of the mission and size of an aircraft and the choice of the appropriate propulsion system, structure, and materials—in short, the whole broad field of aeronautics. Most of these topics are beyond our purposes; however, some aspects of stability and propulsion must be understood. Propulsion is indeed a field unto itself: engine design, fuels and chemistry, compressor characteristics, and other facets are involved. Fortunately, all these topics are covered in simple terms in some of the works listed in Appendix 4. The same is true for discussions of airplane performance such as starting, cruising, and landing, which will also be omitted here. However, a quick discussion of some of the fundamentals of stability, control, and propulsion must round out our previous chapters. We can then proceed to the fascinating aerodynamics of high speeds and the social impact and future of aviation.

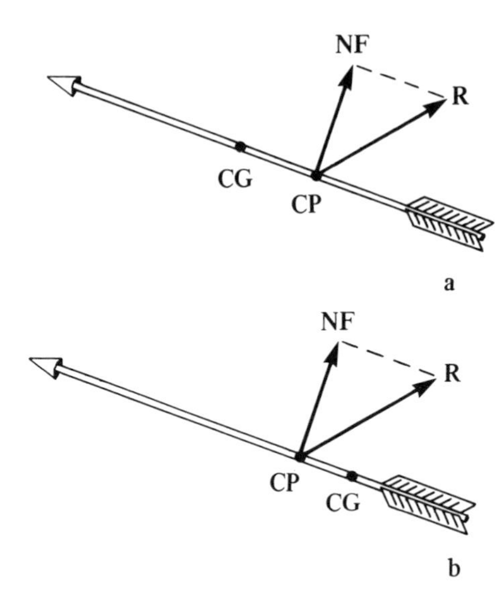

Figure 9.1 *An arrow in flight. (a) Aerodynamically stable flight. (b) Unstable flight.*

We learned that the sum of the aerodynamic forces acting on an airplane yields a resultant air force, R, a vector shown in Figures 1.1 and 8.2. The origin of this vector lies within the aircraft at a point called the *center of pressure,* abbreviated CP. This imaginary point—also called the aerodynamic center—can be viewed as the exact location at which the force R engages the aircraft. The aircraft also has a *center of gravity,* abbreviated CG. If we suspend the airplane by its CG in still air, it will be balanced with respect to gravity and remain motionless irrespective of its orientation. Get an idea of this phenomenon in the simpler situation of balancing a stick on your finger. You can find the CG by trial and error, choosing different support points. The stability of flight—or the lack of it—is closely tied to the relative locations of the CP and the CG.

Let us first study the stability of the much simpler flight of an arrow. We shall simplify even further by restricting the turning motion of the arrow to a single one of the three axes of flight, the pitch—or up-and-down motion (Figure 1.3). Immediately after it is shot from a bow, the arrow displays an angle of attack with respect to the flight direction as seen in Figure 9.1a. The CP is located *aft* of (in flight direction behind) the CG. We must now introduce a new force component closely related to lift and drag. This force is called the *normal force* (NF), and it acts at a right angle to the arrow, or to the longitudinal axis of an aircraft. The normal force produces a *moment.* A moment is defined as the turning effect or torque. For an aircraft, the moment is given as the product of the NF and the distance between the CP and the CG, in analogy to the flying arrow shown in Figure 9.1. Consequently, the dimension of the turning moment is defined by force times length.

Aerodynamic moments are produced by the deflection of control surfaces such as the rudder shown in Figure A2.1, which changes the direction in which an aircraft is flying. Exerting a *given* moment to change the course of the airplane, a smaller force is needed if it is applied farther away from the CG. In turn, for a variable normal force (NF in Figure 9.1), this force decreases to produce a certain moment with a decreasing distance between the CP and the CG. Analogous to lift and drag coefficients (Chapter 6 and Appendix A2.3), we can introduce a dimensionless *moment coefficient* defined by

$$C_M = \frac{\text{moment}}{qS\bar{c}}.$$

Recall that q stands for the dynamic pressure and S is a characteristic area. In order to render the moment coefficient dimensionless, we must divide the moment by a length as well as by qS. The length is usually taken as the *mean* chord, $\bar{c}$, computed by dividing the wing area by the wingspan, the distance between the two wing tips. The length $\bar{c}$ is used also to calculate flight Reynolds numbers of aircraft, like those shown in Figure 6.3. The mean chord has typical values of about 0.2, 1.4, and 8.6 m for the wandering albatross, a hang glider, and a Boeing 747 airplane, respectively.

Common sense tells us that the arrow in Figure 9.1a at a larger angle of attack will lower its tip because the normal force (NF) acts on the CP located behind the CG. By simple inertia the arrow will keep on moving, with its tip dropping below the $\alpha = 0°$ position. At that stage the moment will start to act in the opposite direction, but after a few oscillations the arrow will assume *stable* flight, pointing its tip continually in the wind. This final attitude exemplifies *stable equilibrium;* the arrow has assumed a permanent equilibrium position. The sequence of events leading to stable equilibrium demonstrates *dynamic equilibrium,* given by a number of oscillations about the ultimate stable flight of the arrow. The dynamic adjustment to stable equilibrium is a time-dependent process during which the amplitude of the oscillations—the departure from straight flight—is successively damped.

In Figure 9.1b we have moved the CG far back by putting, for example, a lead weight at the end of the arrow. The CG is now located *behind* the CP; their relative positions on the shaft have been reversed. The tip will now be turned *up* by the moment, and the arrow will tumble. Flight becomes *unstable.* In the last chapter we faced an identical dilemma with the rifle projectile, whose CP was located far ahead of the CG, even ahead of the body itself. In this instance, stability had to be obtained by an entirely nonaerodynamic mechanism, the gyroscopic stabilization achieved by spinning the projectile. Our arrow is aerodynamically stabilized to fly straight by feathering the shaft at its end. This scheme is copied in many rockets and missiles, which have winglike surfaces that produce lift (and drag) to ensure stable flight. A stick is not stable aerodynamically by itself. The spear or javelin used in hunting, war, and the Olympic Games flies stably only when it has a weight such as a steel tip at its head.

Another example to further our understanding of stability comes from the field of mechanics. Take a bowl and let go of a marble near its rim. The marble rolls down toward the bottom of the bowl, overshoots it, returns, and, after a few oscillations, rests at the bottom. As with the arrow, its oscillations about the final position were damped; stable equilibrium was established at the end of the process. If you give the marble a little push, after a while it will again end up in the same stable place. Now turn the bowl upside down and carefully balance the marble on top. A little push suffices to make it roll down the side; of course, it does not return to the top. Its initial position defined an *unstable equilibrium.* A third condition exists, called *neutral equilibrium.* Here our marble lies on a flat surface; if we move it to another location, nothing happens. We achieved neutral equilibrium (with respect to gravity only) when we suspended our airplane or arrow at its center of gravity. In aerodynamics this situation arises if the CG and the CP coincide, defining a most precarious flight condition for the pilot.

In the simplest terms, stable flight exists if the center of pressure is located aft of the center of gravity. The pilot lets go of all controls, and in calm weather the aircraft keeps flying on its set course. If the air force acted at a location ahead of the CG (i.e., the CG would be aft of the CP), the plane would tumble, just like our arrow in Figure 9.1b. In such a situation it becomes tricky indeed to keep the plane on a straight course by operating the control surfaces shown in Figure A2.1. But the plane *can* be flown in a fundamentally unstable mode. We saw this in our discussion of technical details of the earliest airplanes designed and built by the Wright brothers (Chapter 1). The same is true for many of the early airplanes, including those used by the military in the First World War. Even today large ballistic missiles fly in this fashion, with their control systems operated by computer-activated mechanisms. Computers of course are not plagued by physical sensations; they are not subject to feeling like a pilot who is flying "by the seat of his pants."

Let us now turn to the aerodynamic function of the control surfaces. When we wrote the definition of the moment coefficient, we did not single out a particular moment. An actual aircraft has in fact six *degrees of freedom,* or possible motions. The plane flies in three-dimensional space (see the remarks on kinematics in Chapter 5), and it can rotate about its three axes (Figure 1.3). In turn we must deal with six moments. Add to this fact the complication that all moments—just like the forces we discussed before—are a function of flight speed, angle of attack, and the general attitude of the aircraft with respect to its flight path, and you get an idea of the difficulty of controlling flight. In addition, different positions of the airplane in space lead to different exposures of the entire structure to the wind. Variable values of frontal area effectively alter the shape of the craft. More of the fuselage is exposed to the wind in a turn, for example, than in straight flight. This again alters all forces and moments. Consequently, a complex interplay of all degrees of freedom of flight affects stability, a topic whose details we must leave to the technical literature.

Airplanes are ordinarily designed to be stable in level flight. However, an aircraft is constantly subject to a variety of unsettling events. One of these arises from the uneven aerodynamic forces exerted even in perfectly calm air by mechanical asymmetries of external shape. These are *trimmed*—corrected—by minute, permanent adjustments of parts of the control surfaces. The basic functions of the elements shown in Figure A2.1 are nonetheless simple to understand. At the ends of the horizontal and vertical stabilizers (fixed, winglike panels) are hinged movable surfaces. They are the *elevators* and the *rudder,* used respectively to move up and down (pitch) or sideways (yaw). If these small surfaces are deflected, they operate like little wings, producing lift. The angle of deflection determines the lift—or normal force with respect to the airplane—by increasing it until stall sets in as for a wing. The force times the distance to the center of gravity exerts a moment about the pitch and yaw axes, respectively, altering the flight path of the airplane. The third fundamental axis, the roll axis, is controlled by the *ailerons,* movable surfaces at the ends of the wings near the tips. They are actuated in opposite directions to each other to bank the aircraft—that is, to roll it. Such a motion is performed simultaneously with a sideways change of the flight direction, an essential element of controlled flight that, as we have seen, was first practiced by the Wright brothers. Lilienthal controlled the roll axis of his hang glider by dangling his body to shift the center of gravity of the flying machine sideways. But the activation of roll by control surfaces was first applied by the Wright brothers.

To reduce the forces needed to actuate the control surfaces hydraulically or electrically, the moments about the axes must be reduced. This can be accomplished by lowering the inherent stability of the craft, a situation in which the distance between the CG and the CP plays the dominant role. The multitude of movable surfaces in Figure A2.1 all apply to control. They include the slats and flaps that change the lifting characteristics of the wing, and even the spoilers that you can see from your window in an airliner moving slightly up and down in flight and more noticeably during landing. The distance of all the movable surfaces from the center of the aircraft, where the CP and the CG are close to each other, is important to their action. Cruder controls are achieved by the surfaces far from the center, while small adjustments to the aircraft's attitude are accomplished by the closer surfaces.

In principle, as I said, an aircraft can also be operated in the neutral or unstable mode. This is difficult for a human pilot, but an automatic pilot—that is, a computer monitoring the multitude of instruments indicating altitude, speed, location of the horizon, and so forth—can quickly solve the complex equations involved and derive the proper values of control-surface operation to keep the aircraft on course. This black box—the automatic pilot—has now reached a high degree of perfection: it can start and land a plane, react quickly in turbulent weather, keep an assigned course, and perform most of the functions of a pilot (or pilots). It can readily fly an unstable airplane, since it is not prone to the natural anxieties of a pilot, whose feeling of security is coupled to stable flight. Nothing, however, is completely foreseeable. The passenger on a commercial airliner still expects experienced human beings to occupy the cockpit, remaining alert even during long flights and taking over with their human brains and experience if the need arises.

A last remark on control concerns the complicated interplay of the motion of an aircraft and its structure. All the components attached to the fuselage are subject to the aerodynamic forces discussed. For example, the wings move up and down. The differentiated motion of these components in relation to their structural strength is studied in the field of *aeroelasticity,* a specialized branch of aircraft design that will not concern us here. We will simply note that when you see the wing tips move up and down in turbulent weather, you can rest assured that they have been designed to take a beating.

9.2 Propulsion: From Propellers to Rockets

Powered aircraft need engines to provide thrust to propel them. Thrust is defined as the force that pushes an aircraft through the air. We saw in the force diagram (Figure 1.1) that when the aircraft is in level flight at cruising speed, the horizontal and vertical forces cancel each other. The thrust equals the drag, and both have the dimension of force (F). You recall our discussion in Chapters 1 and 8.4 of the seemingly puzzling statement that no net force acts on the aircraft in steady level flight. Moreover, two of the laws of mechanics that we have already discussed govern aircraft propulsion, no matter what system we choose to generate thrust, ranging from propellers driven by rubber bands to rockets. We expressed Newton's law of force in terms of rate of change of momentum, mU (mass times velocity). Force is equal to the rate of change of momentum with time. All systems of propulsion impart momentum to the oncoming air. A propeller—recall the hovering hummingbird—speeds up the air ahead of it. Consider a model with a propeller in a wind tunnel. The air in the test section is blown against the propeller. The propeller is driven by a motor that imparts kinetic energy to the air, speeding up the incoming air. The resulting increase in momentum produces the force of the thrust. The identical process happens with an airplane in flight since the plane's airspeed in still air is identical to the airspeed in the wind tunnel. A jet turbine burns liquid fuel, and the hot gas of the burnt fuel increases the volume of the air and accelerates the flow leaving the turbine. Here the force is produced by a change of momentum that increases both mass *and* velocity. Finally, a rocket, the only power plant that can function without air—that is, the oxygen in the air—in outer space, carries its own supply of the oxygen needed for combustion of its fuel. It spews out a high-speed jet of hot gases, producing the force of thrust.

Once the thrust is provided by any of these methods, Newton's third law takes over. Action equals reaction: a force equal to that produced by the rate of change of momentum, but opposite in direction, pushes an aircraft forward through the air. Thus Newton's second and third laws, which were well understood in the seventeenth century, dominate aircraft and rocket propulsion in the twentieth century. From balloons (Archimedes' law of buoyancy) and the first Wright *Flyer* to the Apollo flights putting men on the moon and the space shuttle, the fundamental laws of mechanics control the motion of flying machines.

The internal-combustion engine powered heavier-than-air craft exclusively until the waning months of the Second World War, when jet turbines that had been developed independently in England and Germany made their appearance. A German fighter plane, the ME 262, was the first operational aircraft powered by a turbojet engine. Extensive concentration on the further development of such turbines led to the jet age of commercial flight, starting in the late 1950s. At that time, propulsion of airliners by gas turbines was introduced on a large scale.

The power of piston engines for airplanes—internal-combustion engines like those of automobiles—ranged from the 12 horsepower (9 kW) of the first Wright *Flyer* to over 2,000 HP (1,500 kW) at the end of their heyday. The most powerful engines of this sort featured a radial arrangement of many cylinders enclosed in a cowling, a metal envelope that drastically reduced the air drag of the engine. Such designs evolved soon after the First World War. Airspeeds of over 500 mph (800 km/h) were achieved with racing planes, only a little less than the flight speeds of current jet airliners. Piston engines also reached high power outputs per unit weight. The Wright brothers had been forced to build their own light engine because the existing automobile engines were much too heavy. However, their power plant, although unique in their time, weighed ten times more per horsepower produced than modern aircraft piston engines. Even higher power/weight ratios are attained by jet turbines, as we shall see, but piston engines are here to stay for some time in many different applications.

Piston engines translate their power into thrust via a propeller. Cayley is said to have designed a propeller-driven model, which was probably not built. His propeller was made of feathers, and he planned to power it by a twisted elastic band. Later in the nineteenth century, propeller-driven model planes with rubber-band motors were flown successfully by a number of aeronautical enthusiasts. Paddles and helical screws, like those envisioned by Leonardo da Vinci, had been proposed even earlier. Hand-operated propellers were attached to balloons in attempts to render them independent of the vagaries of the wind. We discussed some of these inventions in Chapter 1. But again, the first really successful application of this system was the

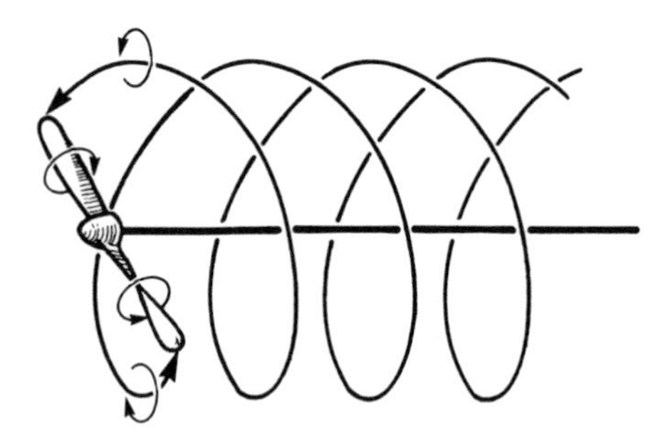

Figure 9.2. *The flow about a propeller at a given instant. The circulation around the propeller and the spiral path of the tip vortices are shown.*

two counter-rotating propellers designed by the Wright brothers. The large size of these propellers permitted a low rate of turning of the drive shaft of their engine and chain drive.

Modern propellers have cross sections that are based on those of airfoils, making them like wings mounted on a turning shaft. Like wings, propeller blades produce lift, drag induced by lift—remember the wing-tip vortices—drag due to friction, and separation. The positioning of the blades is designed to create a lift component in the direction of flight that pulls the aircraft forward. To better understand how a propeller works, we can view it as a disk with a great many blades. In the steady state (see Chapter 5), the air flow approaches the disk in a continuous flow. Increasing the pressure and speed across the disk increases the momentum and produces thrust. The pressure difference is like that on a wing: the front of the propeller has lower pressure than its back. You can also view the propeller as producing a suction effect in front at a right angle to the direction of flight.

If the propeller operates at a fixed number of revolutions per unit time, it has a fixed *angular velocity.* Each point from the hub to the tip traverses a given angle in the same time, just like the points on a wheel of a car that are located on a straight line drawn from the hub to the rim. While the propeller turns, the aircraft moves forward and the flow shown in Figure 9.2 emerges. However, while the angular velocity is constant, the *speed* of different sections of the propeller blades varies, increasing from the hub to the tip. For this reason the blades must be twisted—that is, the angle of attack changed locally—to optimize the lift, or, more exactly, thrust, at each location. Circulation acts in principle as it did for the wing, but its strength also varies along the blade. The general direction of the circulation is indicated in Figure 9.2, in which the helical path of the tip vortices, resembling a corkscrew, is also shown. The angle of the propeller blades is adjustable on many transport aircraft to retain thrust efficiency under different flight conditions. In case of an engine failure, the propeller can be "feathered," turning the blades into the wind to reduce the drag. The propeller action is based on the theory of lift on wing sections; it is not the action of a drill eating its way into the air.

It is now obvious that with propellers we are dealing with a most complex flow system. Because the rotating blades run into portions of their own wakes, the lift distribution is uneven. However, there remains a net component of thrust, the force that pulls the airplane through the air. We shall see later that propellers have a great future. In turboprop mode, strange-looking new propeller designs—some shaped like sickles—may provide the most economical air transportation of all systems, ensuring a continuing life for the oldest form of aircraft propulsion.

We now turn to the *gas turbine,* the power plant of most commercial airliners today. Gas turbines were developed long before their application in aeronautics; they were used to generate electricity with natural gas or coal gas as fuel. Modifications of the basic principles of turbines leading to the *turbojet* that powers aircraft appeared much later. Like the propeller and the piston engine, the turbojet is restricted to flight in the earth's atmosphere, since it requires the oxygen in the air. Flight at the outer edge of the atmosphere and in deep space needs a propulsion system that carries its own oxygen in addition to fuel; here the *rocket* takes over.

The development of turbojets for aircraft was also made possible by other existing technologies besides those involving gas turbines. Much of the internal machinery of turbojets, such as compressors and turbines, benefited from years of research and use in other applications. Axial and centrifugal compressors—in contrast to compressors employing pistons in cylinders—had been applied to move gases in pipelines, operate turbochargers to increase the pressure in automobile carburetors, and do related tasks. The steam turbine—the successor of the piston steam engine—was invented in the nineteenth century to produce power by an expansion of high-pressure steam. Even today steam turbines drive the generators that produce electricity in power plants, regardless of the heat source for the boilers. While the shapes of turbine wheels—the impellers—and related aerodynamic features were applied to the design of turbojets, the weight of stationary power-plant equipment required novel designs for lightweight turbines suitable for the propulsion of aircraft.

The development of different versions of gas turbines to power airplanes is relatively recent. We noted that the first production airplane equipped with such a power plant flew at the end of World War II. Although there was much to learn from past experience, there was also a great deal of new aerodynamics of internal flow systems associated with the design of turbine wheels, nozzles, and the like to handle the flow in jet turbines for flight. This new aerodynamics is in turn intimately linked to research in materials that can withstand the heat of the combustion gases and minimize structural failures of the rapidly rotating machinery, specialized investigations that are far afield from our objectives. But a brief look at the way these power plants work is needed to round out a discussion of airplanes.*

A much-simplified cross section of a turbojet is show in Figure 9.3. The basic method of operation applies equally to the many modifications of jet turbines for specific applications, such as driving a propeller rather than using the exhaust jet for propulsion. The major mechanical components inside the engine consist of entry and exhaust nozzles, an axial compressor and a turbine, both mounted on a

*For the history of turbojet development, see, e.g., the books by Anderson and Shevell cited in Appendix 4.

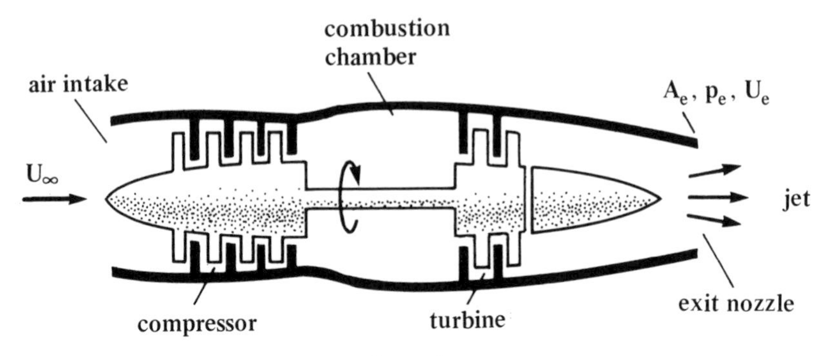

Figure 9.3 Schematic drawing of a turbojet engine.

single shaft, and a combustion chamber located between the two.

The axial compressor increases the air pressure between the intake and the combustion chamber. It has rather short, propeller-like vanes rotating on the large-diameter central shaft. Interspersed with the moving parts are fixed vanes called stators that are mounted on the outer wall. Several rows—called stages—of turning and fixed vanes are used. In each stage the pressure is increased by a fraction of the overall pressure increase across the compressor. In the axial arrangement, the air flow in the compression cycle moves largely parallel to the shaft.

The turbine is of a different design. It expands the air—here the pressure drops through each stage—and rotor and stator vanes are longer; the air and the products of combustion wind their way sideways and lengthwise during the expansion. In fact, the turbine of a jet engine looks much like a steam turbine.

The aerodynamics of compressors and turbines is a difficult field, again somewhat separate from our interests. For a power plant like that shown in Figure 9.3, fluid mechanics research and testing requires the placement of probes to measure pressure, airspeed, and air direction on the rapidly turning wheels. Electrical connections must be led to the outside via the turning shaft. It is difficult to photograph the flow because of the obstruction by the elements under investigation, and for new designs simplified models of two-dimensional flows are looked at first. Special laboratories exist that are wholly devoted to this research.

The entire gas turbine is housed in a streamlined enclosure designed for low drag. This engine nacelle (see Figure A2.1) may be mounted on pylons under the wing or on the side of the rear of the fuselage, or it may be integrated with the vertical stabilizer.

How does the aircraft gas turbine, the jet engine of Figure 9.3, work? (Here we omit turboprops, jet engines that drive a propeller via gears, not unlike the gas turbines that long ago used to drive electric generators.) Let us apply our wind-tunnel coordinate system, putting the engine in a wind tunnel and restricting ourselves to operation at airspeeds below the speed of sound. This restriction applies to the great majority of commercial airplanes currently in use (see Chapters 2 and 10). Special wind tunnels exist in which a full-size engine can be mounted. The engine is operated in the tunnel in order to study the flow through and around it, including entering air and the exhaust jet. The entry speed at the air intake is close to the wind-tunnel test-section speed U_∞. (In free flight, the intake speed is practically equal to the flight speed.) The cross-sectional area of the intake widens beyond the entry face. We know from Chapter 5 that the conservation of mass requires that the flow speed decrease if the area increases. From Bernoulli's equation we find that the pressure increases; at the face of the compressor its value is nearly that of the stagnation pressure p_0. This diverging intake duct is called a *diffuser.* The same aerodynamic configuration of a gradually widening cross-sectional area is used in wind tunnels to reduce the airspeed in the duct connecting the test section to the fan that drives the wind tunnel. In both cases, abrupt changes of area are avoided, to keep the flow attached to the walls and prevent a loss of energy.

The stopping of the air, or ram effect, together with the drag of the outer shape (of a certain pressure distribution, friction, and separation), contributes to the drag of the jet turbine and in turn to the drag of the aircraft. These external components of drag are called *parasite* drag. Next, the air pressure is further increased beyond the diffuser in the axial compressor. It therefore enters the combustion chamber at a higher pressure than p_0. Here liquid jet fuel is injected via spray nozzles (not shown in Figure 9.3), ignited, and burned. The oxygen required for combustion is provided by the oxygen in the air (see Chapter 4). The combustion chamber is now filled at high pressure with a mixture of air and products of combustion at a temperature of over 1,000°C (1,832°F). The problems of materials and structure for the combustion chamber and the turbine beyond it are now obvious. The turbine is driven by the hot gases, and in the expansion through the stages of the turbine, shaft power is produced. With the turbine mounted on the same shaft as the compressor, the turbine drives the compressor. But there is energy left over in the form of kinetic energy of the hot gases. These gases are expelled at high speed at the exit nozzle, providing the thrust to push the airplane through the air. Here Newton's third law (action equals reaction) takes over, just as it applied to the propeller.

The exhaust jet is formed in the exit nozzle, a contracting duct. Decrease of cross-sectional area implies an increase in the airspeed and a drop in pressure. We observe the opposite effect discussed for the air intake or diffuser. (The two processes are illustrated in Figure 5.5.) The pressure in the exhaust jet, however, is still above that of the ambient air, but soon it assumes that pressure.*

It remains for us to compute the force of the thrust (see Figure 1.1) of a jet turbine acting to push an airplane. The thrust of a turbojet has two components; its major part is produced by the difference in momentum of the air at the intake and the air (including the products of combustion) in the exhaust jet leaving the exit nozzle. Recalling our earlier remarks on changes in momentum induced by a propeller, we must now determine the rate of change of momentum for the jet turbine. Here we have an initial flow speed, U_∞, entering the intake (for symbols in the following, consult Figure 9.3). This is either the wind-tunnel or the flight speed. At the exit nozzle, the exhaust has the flow velocity U_e. We saw that the turbojet operates because the latter speed is higher; we have $U_e > U_\infty$.

But there is also a change of mass to be considered to determine the change of momentum, since fuel was added in the combustion chamber. The fuel mass added per unit time is small in relation to the air gobbled up at the intake. Typically we find a mass fraction of fuel in the air of about 5%. Consequently the mass rates entering and leaving the turbine are practically equal, and they are virtually given by that of the entering air. However, as we noted from the comparison of density of liquids—here the fuel prior to ignition—and gases (see Table 3.1), the gaseous burnt fuel has a much lower density, resulting in a much larger volume than it had in the liquid state at the point of injection into the combustion chamber. The large volume increase—by a factor of the order of 1,000—increases the pressure and is part of the effect of speeding up the gases. Consequently, the increase of momentum between the air intake and the exit nozzle of a turbojet is given primarily by the increased flow speed at the exit.

There is an additional effect that adds to thrust. At the exit section, A_e, where the high-speed exhaust jet leaves the engine, the difference between the jet's pressure, p_e, and the ambient pressure, p_∞, must be considered. In fact, the jet's pressure is much higher than that of the surrounding air, resulting in a contribution to the push given to the airplane by the turbine. Taking momentum and pressure effects into account, we can now write for T, the thrust,

$$T = (dm/dt)\,(U_e - U_\infty) + (p_e - p_\infty)A_e.$$

---

*At high altitudes the jet turbine leaves a condensation trail ("contrail") in its wake. Among the products of combustion in the jet we find water vapor (H_2O in the form of steam) entering the cold ambient air (see Table 4.1). With the air temperature being well below the freezing point of water, ice particles are formed that mark the flight path for long distances in a clear sky.

Here we have used the symbol dm/dt for the mass discharge per unit time. This is to be multiplied by the velocity difference to obtain the change of the momentum from entry to exit. The first term in the equation shows that the thrust increases if the difference between exit and entrance speeds increases. This increase is a linear one: twice the speed difference gives twice the thrust. The speed difference is of course positive, since the speed at the exit is much higher than that at the entry. This result could also have been deduced by common sense. With exit pressure above ambient pressure, the second term also adds to the thrust. In the thrust equation we ignore the aerodynamic parasite drag force of the turbojet and its engine nacelle. This value must be subtracted to compute the net effective thrust available to push an airplane.

It remains to determine, recalling our discussions of dimensions and units (see Chapter 6 and Appendix 2.2), if the dimensions of the terms on the right-hand side of the equation in fact result in that of the force of the thrust T. This problem is most readily solved for the second term. With $p(F/L^2)$ and $A(L^2)$, the product of the two gives the required force F. It is instructive for the reader to use Table A2.2 to verify that the first term in the thrust equation has the identical dimension. Typically the force unit pound (*lb*) is used to give a quantitative answer for thrust.

For comparisons of jet turbines with piston engines, it is interesting to compute the power of each device. The product of flight speed times thrust gives us this answer, again recalling Table A2.2. With the foregoing equation for thrust and our knowledge of dimensions and units,* we are now able to determine thrust and power, provided we know the flight speed, flight altitude (to determine the ambient pressure from Table 4.2), and the characteristics of the jet turbine. It is noteworthy that the thrust and power of an advanced technological device such as a turbojet can be calculated from simple laws of mechanics using simple algebra. In our discussions of the thrust and power of propellers and turbojets, we have, of course, implicitly assumed steady and one-dimensional flow (see Chapter 5). Both assumptions are valid approximations for cruising flight at constant speed and altitude.

The remarkably simple thrust equation tells us that thrust is improved by increasing the velocity difference between intake and exit, and/or pushing more mass through the system. I repeat that common sense tells us the same thing without any equation. If you increase the

---

*Force times length gives us the work (or energy). Work per unit time expresses power, P. With speed as length over time, $P = (FL)/T = FU$, as we noted before in our discussion of automobile power.

water flow in a garden hose, it pushes harder in the opposite direction. With everything else being the same, thrust also improves at higher altitude with lower ambient pressure. But looking at our discussion, a puzzling element remains. Are we emulating the legendary Baron von Münchhausen, who extricated himself from a swamp by grabbing his hair and pulling himself up? The power generated by the turbine drives the compressor on the shaft, yet we achieve a net propulsive effect. The designer of a turbojet divides the power produced by expansion of the combustible mixture in the turbine in a certain proportion between the power required to drive the compressor *and* the kinetic energy of the exhaust jet per unit time, the jet power. The choice of pressures produced at the various locations in the turbojet, the efficiency of combustion (a field unto itself), and the structural problems alluded to before all enter into the design of these power plants geared to a suitable distribution of power.

Shifting these elements, we can also dispense with the exhaust jet as the source of propulsion, adding instead a gearbox to the shaft to drive a propeller. This version of the gas turbine is called a *turboprop*. It is an efficient way of powering an airplane. (In a nontechnical way, better efficiency simply means that a unit of weight, say a passenger, can be flown a given distance on less fuel.) Many smaller, short-haul commuter planes make use of this scheme. If maximum airspeed is not especially important, the turboprop can also be used for large transport aircraft covering long distances. Advances in propeller design mentioned before will very soon permit turboprop power plants to fly aircraft at practically the same speed as turbojets, with the turboprop being more efficient.

But let us put down some numbers to help us understand the advances in engine design. We saw that radial piston engines driving propellers could produce 2,000 HP (1,500 kW). The best engines achieved about 0.9 HP per pound of engine weight. This was an improvement by a factor of ten or so with respect to the first engine of the Wright brothers. A turboprop engine producing the same 2,000 HP achieves 2.5 HP/lb, an improvement of a factor of over 2.5 with respect to the best piston engines. Aeronautics again was the driving force to realize such great advances since the piston days of the 1940s.

A third version of the gas turbine, the *turbofan,* currently powers larger airlines. Here a fan of a *larger* diameter than that of the turbojet is mounted in front of the turbojet on the same shaft. This *centrifugal* fan, which looks like a set of many propeller blades, is what you see on the front of many jet engines when you face the intake. Large turbofan engines have compressor fans of over 2-m (7-ft) diameter. A full-power start at sea level requires that they swallow about 600 m^3 of air per second (21,000 CFM; remember the discharge rate Q of Chapter 5). Large aircraft such as a B-747 jumbo jet have four such engines to push them into the air. This large air mass is split by the fan into an outer stream moving past the turbojet in a concentric annulus—an open duct wrapped around the turbojet—and the air fed into the compressor of the turbojet. The entry of the turbojet now receives air at a pressure above that of the previously discussed ram pressure, and the air is preheated by compression. Beyond that, the workings of the turbojet are exactly as described before. At the exit nozzle the pure air that bypassed the turbojet surrounds the hot exhaust jet, and it has several times the volume of the hot-air stream. Consequently, a larger force or change of momentum is achieved, and the noise of the heated exhaust is reduced.

There are many additional tricks to improve efficiency, reduce jet noise (the scourge of people living near airports), and increase the speed range of gas-turbine engines. A turbojet for the Concorde, for example, must be designed to handle a supersonic airflow at Mach number two entering the intake. Reheaters and afterburners squeeze more jet power out of the combustion gases. The *ramjet,* an exotic device that works only at supersonic flight speeds, improves propulsive efficiency, but an aircraft must have regular jet turbines to get it up to such speeds.*

Finally, we shall look briefly at rocket propulsion. This mode is of marginal interest to aviation, and it is unlikely that it will be applied to commercial aviation in the foreseeable future. However, there is great interest in rockets in the context of military and space applications. The first combination of aircraft and rocket appeared in World War II. Rocket-assisted takeoff permitted military airplanes to use short runways and carry heavy loads. After the war, experimental rocket airplanes carried aloft by transport planes flew with rocket propulsion and returned to the ground as gliders, foreshadowing the space shuttle returning from the outer edge of the atmosphere. In 1947, the Bell XS-1 experimental rocket plane piloted by Captain Charles E. (Chuck) Yeager was the first airplane to break the legendary sound barrier (see Chapter 10) in level flight.

Rockets are powered by liquid or solid propellants, the chemicals that burn to produce thrust. A highly schematic plan for a liquid-propellant rocket is shown in Figure 9.4. This drawing looks much like the rockets envisaged by the early pioneers Goddard, Tsiolkovsky, and Oberth, working respectively in the United States, Russia, and the German enclave in Romania. A combustion chamber is fed with a liquid fuel such as alcohol and an oxidizer. The latter could be liquid oxygen, as used in World War II by the German V2. This substance is difficult to handle because of its extremely low tem-

---

*Much of the material discussed here was taken from the books by Shevell and Anderson (see Appendix 4), which include descriptions of these and other interesting details of propulsion from propellers to ramjets.

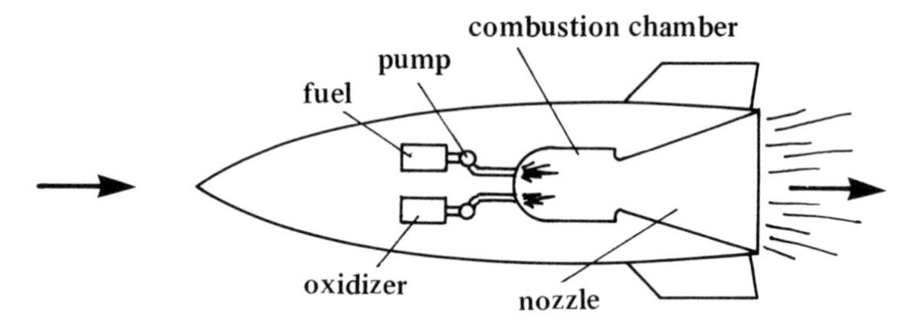

Figure 9.4. Schematic drawing of a liquid-propellant rocket.

perature (-185°C, or -301°F, at atmospheric pressure). But there are liquids whose molecules are rich in oxygen that can be stored in tanks at atmospheric conditions. From there the two liquids are pressurized and injected into the combustion chamber, where they are ignited.

The products of combustion flow at high pressure and temperature into an exhaust nozzle specially shaped to produce a jet at supersonic speeds (see Chapter 10). Such a nozzle must be designed according to laws that take the compressibility of gases into account and that are applicable to flows that have speeds faster than the speed of sound. These laws are extended versions of the flow equations that we developed in Chapter 5. The resulting nozzle looks like that shown in Figure 9.4. Past the combustion chamber, the cross-sectional area decreases to a minimum—a throat—at which the hot gases reach the speed of sound. Beyond the throat is an expanding section where the gases move at increasing supersonic speeds. At the exit the supersonic jet emerges and spreads out. This behavior of gas flow in a converging-diverging passage, a nozzle, is exactly opposite to that shown in Figure 5.5; it is a result of the flow equations of compressible flow, where the density becomes a variable.

Returning to our thrust equation, we note that it can be readily adapted to the rocket. No air is taken in, and thus the entrance speed, U_∞, is zero. The momentum results from the product of the rate of mass exhausted times the jet speed. The second term, as before, demonstrates that the thrust increases by lowering the external pressure. The rocket is the only propulsion system of those we have discussed that can work in space. In this environment the ambient pressure $p_\infty = 0$. Consequently, the maximum thrust is attained, everything else being the same, as we see by our equation. Again, from known thrust and the speed of the rocket, we can compute the power produced to propel the device. For a huge rocket like the Saturn, a truly awesome power is developed. This high value was needed to achieve the speed required to escape the gravitational pull of the earth—about 11 km/s (7 mi/s)—and to lift the heavy load of the Apollo spacecraft beyond the earth's atmosphere on its lunar mission.

Chapter Ten

Toward High Speed: Supersonic and Hypersonic Flight

Robins (1707–51) we must suppose never imagined in his wildest dreams that ordinary people would one day fly as fast as the shot from a gun, but his work was nevertheless the origin of modern aerodynamics. He showed how subtle and how brutal could be the forces of the air and, in particular, how the speed of sound would be a barrier to progress.

CHARLES BURNET
(THREE CENTURIES TO CONCORDE, 1979)

10.1 Pushing the Speed of Sound: Flight at M = 1?

Just like that of trains and automobiles, the speed of aircraft has increased with time. Indeed, aircraft have raced ahead to make the most substantial progress in the shortest time. This is shown in Figure 10.1, where an exponential increase of cruising speed with time up to about 1960 is apparent. The curve is a rough average drawn through points that designate the cruising speed of well-known airliners marked at the time of their initial service. We see, of course, what the experimentalist calls "scatter"—deviations from the mean—because no smooth curve can possibly be expected to apply to such a human endeavor.

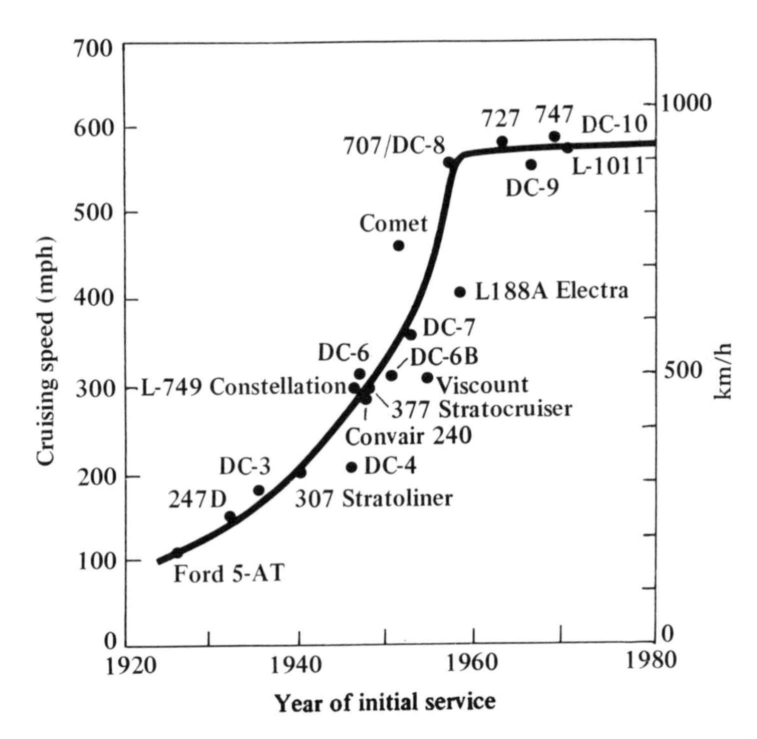

Figure 10.1. Development of the speed of subsonic transport aircraft from 1920 to 1980.

Some of the deviations are of particular interest. The Douglas DC-7, introduced as the last large aircraft with four internal-combustion engines (the Viscount and the Lockheed Electra had turboprops; see Chapter 9.2), was preceded by the British Comet, the first four-engine, pure jet airliner. The dramatic speed increase of jets is apparent. The Comet developed structural failures of the fuselage leading to explosive accidents that caused its withdrawal from the market, but the age of jetliners was here, with the Boeing 707 and the Douglas DC-8 entering regular service shortly before 1960.

You can see a discontinuity in the slope of our speed-time graph: the curve flattens, and its trend has not changed to this day—sixteen years after the end of the time scale of the graph—even though there are several new transport planes and improved versions of existing aircraft flying. Typically, the large commercial airliners introduced by Boeing during the years 1982 to 1995 have normal cruising speeds of about 530 to 555 miles per hour. The same is true for similar airplanes of other manufacturers such as McDonnell-Douglas and the European Airbus consortium. What happened? Why did the curve not proceed on its exponential course?* Why do all these airplanes fly at the same speed? This story must now be told.

Our eighteenth-century artillerist Robins (see Chapter 1.4, the epigraph to this chapter, and Appendix 2) had observed that he could increase the range of his cannonballs by increasing the powder charge, but he also noted to his surprise that beyond a certain charge, no further significant increase resulted. In modern terms, the drag coefficient of cannonballs must have increased significantly as Robins's muzzle velocities approached the speed of sound. Why is this so?

Recall that the Mach number is defined by division of flight speed (or the flow speed in a wind tunnel) by the speed of sound; approaching that speed is therefore equivalent to approaching M = 1. The value of Mach number one is the watershed between subsonic and supersonic flows, two types of flow that behave in very different ways indeed, as we shall see.

Now to the meaning of the speed of sound. If you throw a pebble in a quiet pond, a wave propagates outward in an ever-widening circle. The little wave proceeds, while the local fluid elements of the water do not move with it. They follow an up-and-down orbital motion. A bodysurfer is aware of that internal movement of the fluid elements. The propagating wave itself is a *transversal* wave. Its crests and valleys—its variations of amplitude—move at a right angle to the direction of propagation. Electromagnetic waves are of the same type. Wavelength and frequency are terms that are well known to radio and television owners.

Sound waves, in contrast, are *longitudinal* waves, propagating as a succession of compressions (pressure increases) and expansions (pressure decreases or rarefactions) in the air. Again the fluid, in this case the air, does not move with the sound. It moves locally in the narrow region where it is compressed or expanded by the passing wave. The wavelength is now the distance between the points of maximum compression and expansion. This concept is more difficult to grasp than that of the waves breaking on the beach or produced by the bow of a ship. The frequency of sound, finally, is found by counting the maxima and minima of compression and rarefaction that pass a fixed point in space per unit time.

*Figure 10.1 is a good example to warn engineers not to extrapolate—that is, extend beyond known facts by assuming things will go on as they have in the past. In the 1950s there appeared to be no obvious reason to keep passenger aircraft from flying at ever-increasing speeds. Watch out for predictions that at a certain time there will not be a drop of oil left in the ground, the last tree will have been turned into a newspaper, and so forth.

Take the vibration of the surface of a loudspeaker. If a single note is fed electrically into the speaker, perhaps by a demonstration record or tape, the air directly adjacent to the speaker is made to vibrate back and forth. Depending on the forward or backward stroke, it undergoes a compression or an expansion cycle. The resulting tone is propagated at the speed of sound. The speed of propagation of this note, or, in fact, the speed of propagation of the sound of a gunshot, is *independent of the frequency*. This is a fortunate fact of nature. If it were otherwise, a symphony orchestra would sound terrible, with the high note of the flute reaching a listener at a time different from the low note of the bass.*

At distances such as those between people talking in a room, sound reaches a listener nearly instantaneously. But in a storm you note the time interval elapsing between the flash of the lightning and the arrival of the thunder it causes. In fact, this noise is caused by lightning producing a weak shock wave, another wave that we shall discuss shortly. By remembering the value for the speed of sound and counting or using a stopwatch, you can determine your distance from the lightning.

Finally, we can determine the speed of sound in gases, a, which depends on the nature of the particular gas and its temperature. For air, we find a simple formula

$$a = 20\ (T)^{1/2}\ \text{m/s}.$$

Here the temperature, T, must be the absolute temperature in degrees Kelvin (K) as defined in Table 4.1. Using a calculator and the sea-level temperature of the standard atmosphere of T = 15°C = 288K = 59°F, we find a = 339 m/s = 1,110 ft/s = 1,220 km/h = 758 mph. In the stratosphere where airliners cruise, a = 1,060 km/h = 659 mph. Note from Figure 10.1 that current transport aircraft cruise at speeds *below* that of sound in the stratosphere. They fly at Mach numbers below one, but quite close to it. Robins's observations must therefore apply to them; we are now ready to explain what really happens.

To do so, we must extend our knowledge of ideal (inviscid, without internal friction) flows (see Chapter 5.2). In particular, the

*The good (young!) human ear can hear frequencies ranging from less than 50 to 16,000 hertz (Hz, cycles/s). This leads to a large corresponding range of wavelengths, from 7 m (for the low note) to 0.2 mm. Here we have used the formula that wavelength is equal to the speed of propagation divided by the frequency, an equation that is equally applicable to electromagnetic radiation such as light. (Obviously, for light we must use the speed of light, 300,000 km/s.) The musically minded will recognize that many octaves are encompassed in this range, with the piano displaying about seven of them. Human hearing is a remarkable faculty, responding to a displacement of the eardrum of the order of the diameter of a molecule. Visible light, in contrast, covers less than a single octave from the red to the blue end of the spectrum.

area-speed-pressure relations shown in Figure 5.5 are important in explaining what happens at high speed. Although the figure concerns a converging-diverging duct, flow about a wing behaves similarly. The wing affects the flow just like a duct sliced in half along the centerline. Let the airplane increase speed beyond the range familiar to us. It follows from Bernoulli's equation that the difference between stagnation and static pressures increases accordingly. Recall that this increase is a function of the *square* of the flow speed. At some flow speed the resulting pressure differences measured as a fraction of the ambient pressure will be high enough to change the volume (remember the bicycle pump) and correspondingly the density. *The flow can no longer be treated as incompressible.* The powerful pressure changes in the flow field associated with high speed cause a variety of strange phenomena.

At what point does this start? Compressibility effects—changes in flow parameters that are not precisely predicted by the equations in Chapter 5.2—must be taken into account in aerodynamic design once the flight speed reaches roughly one third of the speed of sound—that is, about 400 km/h (250 mph). Modified equations of motion apply that take variable density into account. But things really get serious at the much higher flight speeds of current jet airliners shown in Figure 10.1. In diving tests of fighter aircraft to attain supersonic speeds that were carried out largely in England after World War II prior to the availability of powerful jet engines, the steep increase in drag observed by Robins was found again. In addition to the increase in drag, stability and control problems arose. The center of pressure shifted sufficiently to make the airplanes unstable, strong vibrations appeared, and in some instances, fatal crashes ensued.

The newly discovered phenomena that dominate such a complicated flow field are pictured in Figure 10.2. Here we see the flow about a Remington rifle bullet moving at the near-sonic speed of $M = 0.95$. Supersonic flow occurs smoothly near the front of the bullet. It is invisible in the photograph. The supersonic zone—faster than the speed of sound—is abruptly terminated by a *shock wave*. Additional waves, including a weak shock preceding the projectile, are visible. The photograph was taken by shining a microsecond (10^{-6} s) flash of light at the right instant at a right angle across the flight path of the bullet. The changes of air density in the compressible flow alter the index of refraction of the air; accordingly, the light rays are deflected. Depending on the variable density, more or less light than initially present is recorded on the photographic plate. Think of the wavering hot air above a radiator, where temperature changes cause the same optical effect to show us details of the rising warm airflow.

The flow about the bullet flying at practically constant speed is steady; the photograph would look identical if taken a moment later. However, the difficult concept of a shock wave is explained more

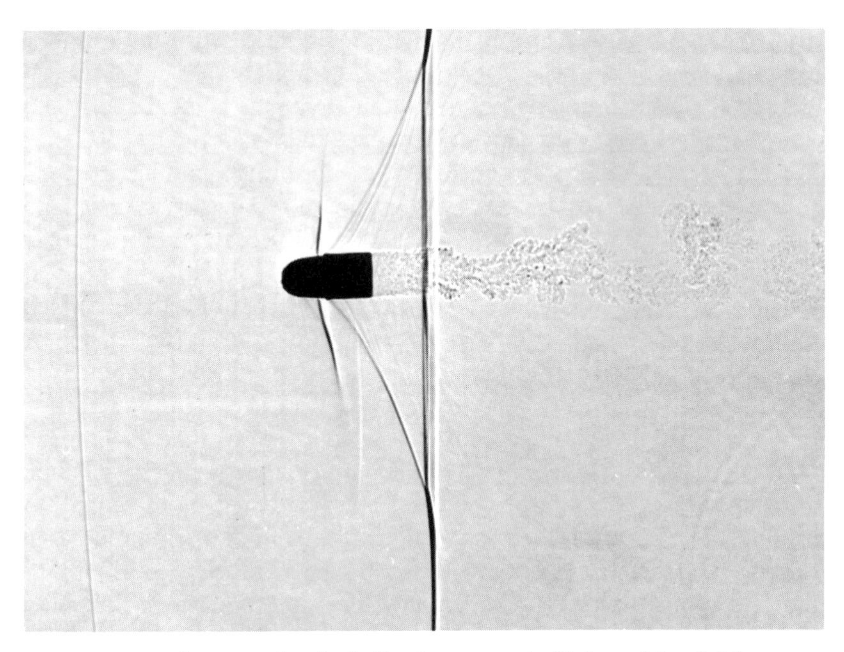

Figure 10.2. Photograph of a bullet in transonic flight at M = 0.95.

readily by reverting for a moment to unsteady flow, the buildup phase of such distinct discontinuities seen in Figure 10.2. Small pressure changes in the flow are transmitted through the air with the speed of sound. Shock waves—discontinuous or abrupt changes of pressure, density, and temperature—show up as fine lines in the photograph. A shock wave is somewhat misnamed because it is really not a wave such as we discussed before. A shock typically appears in explosions, as you may have seen in movies of nuclear detonations. The shock originates by the coalescence of a sequence of small individual pressure pulses, resulting eventually in a strong discontinuity in the flow.

The density changes in compressible flow cause temperature changes; thermodynamics enters aerodynamics. This is a new problem with which we shall deal only in a descriptive manner. A small pressure disturbance propagating in the medium heats the air slightly. In turn, the next pressure bump goes a little faster than the preceding one, since the speed of sound increases with temperature. In this fashion, waves can catch up with each other to increase the ones ahead of the pack. This is shown in Figure 10.3, where a group of unfortunate skiers demonstrates the formation of an unsteady shock wave. The presence of the shock wave is transmitted to the right (or backward), with time running from top to bottom.

Back to steady flow and our topic of flow about a wing. In Figure 10.4 we see an airfoil model mounted on a support on the wall of a wind tunnel to study the flow above its top. The wind blows from left

Figure 10.3. A humorous representation of the formation of a receding shock wave.

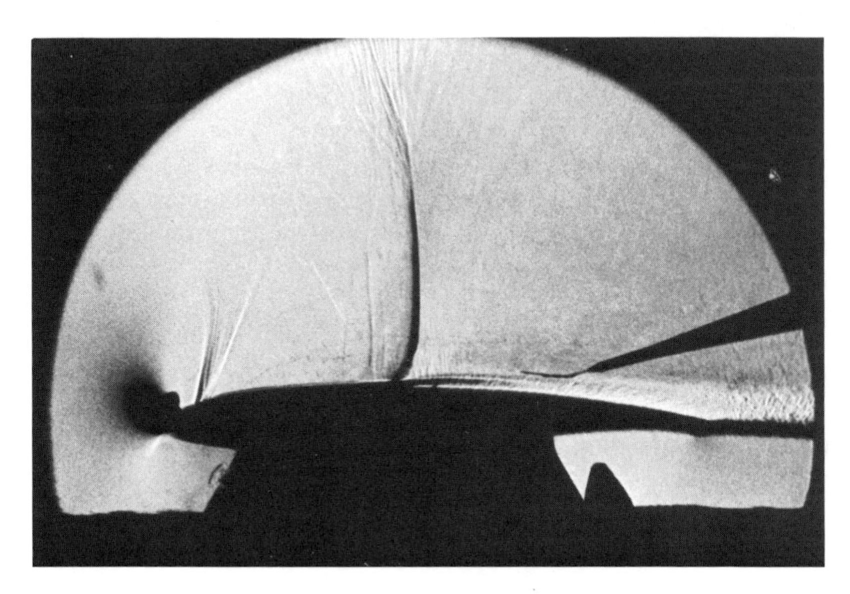

Figure 10.4. An airfoil model (CAST/DOA1) in a Göttingen (Germany) wind tunnel at M = 0.77 and α = 3°. Flow from left to right.

to right, and the wind speed ($M_\infty = 0.76$) is held constant to observe steady flow. The air speeds up over the top of the airfoil for the reasons discussed previously. In front of the maximum thickness of the wing the flow becomes supersonic. This condition is abruptly terminated by a so-called normal shock wave, a shock wave oriented at a right angle to the outer flow direction. In a normal shock, the flow is always changed from supersonic to subsonic speed. A movable Pitot tube is mounted behind the shock to map the flow field. Looking at this picture, which was taken by an optical apparatus similar to that used for Figure 10.2, it is obvious that no smooth flow like that shown in Figure 5.3 exists. Such chaotic behavior increases the drag of the wing. In fact, shock waves dissipate energy much like the internal friction discussed before; this dissipation shows up as a drag. The additional drag component adds substantially to the drag effects discussed in Chapter 7.

In addition to this difficulty, a distorted pressure distribution is measured on the wing, and a curve quite different from that shown in Figure 8.14 arises. This troublesome effect leads to a shift forward of the center of pressure (see Chapter 9.1). It is this problem that led to the instabilities experienced by the pilots of the early diving tests, where the control action was often reversed—up became down and vice versa, with disastrous consequences.*

These experiences, supplemented by wind-tunnel tests and accumulated observations with experimental airplanes, led to the notion of "the sound barrier" or "the sonic wall," an unbreakable barrier to flight. This notion was forever demolished by Captain Charles (Chuck) Yeager's flight on October 14, 1947. Yeager piloted the Bell XS-1 rocket plane that was released at high altitude from a carrier plane. The craft was next accelerated to attain M ~ 1.07 in the first level supersonic flight performed by man. Quite aside from this event, it was long known that artillery shells, high-speed rifles, rockets such as the V2 of World War II, meteorites, and materials propelled by explosions had long mastered the imaginary sonic wall. The only real question was how to do it safely, efficiently, and economically for manned flight.

For aircraft flying near the speed of sound, the relatively new area of aerodynamics dealing with these problems is called *transonic* flow, a term coined by von Kármán. This expression covers the flight range where "mixed" subsonic and supersonic flows occur simultaneously near a moving object. In Figures 10.2 and 10.4, both objects were exposed to a subsonic flow, with supersonic zones appearing

---

*For a discussion of these early experiments, see the book by Charles Burnet cited in Appendix 4.

near the surfaces for the reasons given. Conversely, the term *transonic* applies to objects that move faster than the speed of sound at Mach numbers greater than one, but on which subsonic zones exist locally. The theoretical and experimental treatment of transonic flows is more recent and in fact much more difficult than that of pure subsonic or supersonic regions. Aerodynamicists had to turn to novel solutions to make transonic flight for airliners more manageable.

The first of these solutions was proposed by the German Adolf Busemann, one of the pioneers of gasdynamics. (Gasdynamics is the branch of fluid dynamics dealing with compressible flow, in particular supersonic motion.) In the 1940s Busemann suggested dispensing with the straight wing and using a *swept-back* wing instead. This wing form delays to a higher Mach number the troubles encountered near the speed of sound. *Sweepback,* which also applies to the horizontal and vertical stabilizers, is shown among the multitude of aircraft configurations in Figure A2.3. Our sample airplane in Figure A2.1 has such wings, and the V-shape is typical of all large airliners in operation today. An ultimate version of this wing form—the triangular or *delta* wing—appears on supersonic aircraft such as the Concorde and many military airplanes.

To understand how swept-back wings alleviate some of the transonic problems, a little more background is needed. Pushing a straight wing to higher flight Mach numbers, we find that the local Mach number on the upper, curved surface increases beyond the value of the flight Mach number. This experimental fact is explained by the equations of motion for low-speed flow of Chapter 5. Indeed, as long as flows remain subsonic, the classical equations still describe the phenomena qualitatively. We saw this in Chapter 9.2 in conjunction with the diffuser for the jet turbine. However, we now need modified equations taking the variable density into account to compute the proper values of pressure versus area and the like. At supersonic speed, much of the flow behavior is turned on its head. For example, a diverging nozzle produces an *increase* in Mach number and speed. More about this in the next section. We saw this effect before, looking at a rocket in the last chapter.

As the speed of the airplane increases, we single out two characteristic values of the Mach number that affect the aerodynamics of the wing. The first of these is the *critical* Mach number. It is defined as the subsonic, free-stream Mach number at which sonic flow for the first time appears somewhere on the wing. A second value is known as the *drag-divergence* Mach number; it is the free-stream Mach number at which the drag rises precipitously. This second Mach number follows the critical Mach number very closely; since it is only a little larger, it will not concern us here.

Busemann pointed out that the flow speed that dominates this problem is not the speed of flight but rather the component of that speed that hits the wing at a right angle to the leading edge (see Figure A2.2). Tilting the wing backward—the sweep—reduces this speed with respect to the flight speed. Consequently, the critical Mach number will be increased. An increase in critical Mach number implies that the first appearance of supersonic flow occurs at a higher flight speed for a swept wing than for a straight one. The sweep angle is defined as the angle of the wing with respect to a line perpendicular to the longitudinal axis of the aircraft. Specifically, this angle β (beta) is measured on the 25% chord line, a line positioned one quarter of the chord (see Figure A2.2) from the leading edge.

Calling the Mach number at a right angle to the wing, the component that counts, M_r, we find from elementary geometry

$$M_r = M_\infty \cos \beta.$$

Here, as before, M_∞ is the flight Mach number. It follows that the Mach number component M_r is smaller than the free-stream Mach number. Say that for a given wing, $M_\infty = 0.7$ is the critical Mach number. With a typical sweep angle of 30°, a new, increased critical Mach number is given by 0.7/0.87 = 0.8, since cos 30° = 0.87. A substantial increase of the critical Mach number results for the swept-back wing.

What about a further increase of the sweep angle to make things even better? The sweep affects the streamline pattern above the wing, the circulation is altered, and serious problems arise near the wing root at the fuselage. The vortex flows and the distribution of the lift forces along the wingspan (discussed at the end of Chapter 8) vitiate the beneficial effects of further delay of the critical Mach number. In each instance a compromise—as always—has to be reached among the conflicting behaviors of different aspects of the flow.

A more recent discovery has further eased the problems of transonic flow. Richard Whitcomb of the NASA Langley Laboratory proposed a wing section that includes only a slight curvature at the top of the nearly flat airfoil and a thin trailing edge.* This new shape, the so-called *supercritical* wing section, has now been brought to a high degree of perfection in several laboratories in the United States, United Kingdom, and Germany working independently of each other. The flow about such a wing section is shown schematically in Figure 10.5. The *sonic line,* the location of the change of flow speeds from subsonic to supersonic conditions above the wing, originates practically at the leading edge. An extended supersonic zone is terminated by a weak shock wave. The shock still causes a sufficient increase in pressure to separate the boundary layer. This breakaway (see Chapter 5) is quickly followed by a reattachment of the boundary layer, a phe-

*R. T. Whitcomb, NASA Report No. 1273 (1956).

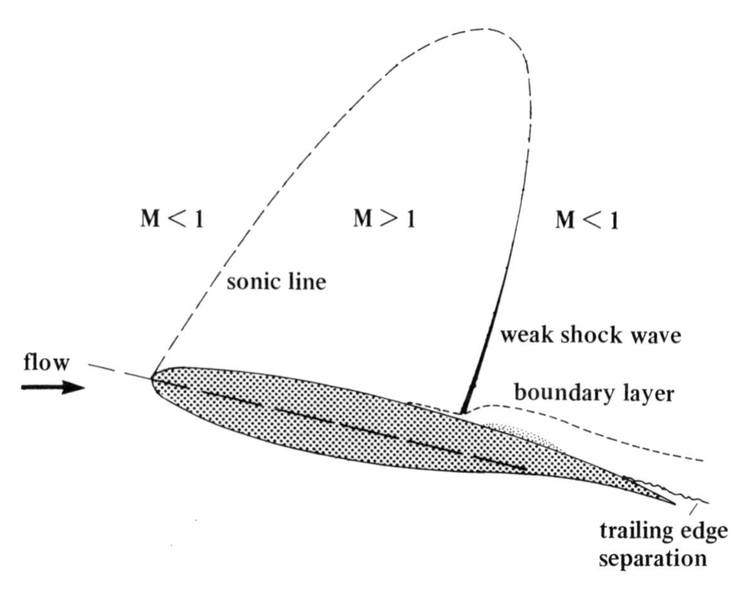

Figure 10.5. A sketch of the shape and flow of a supercritical airfoil designed for low drag at a flight speed slightly below the speed of sound.

nomenon that occurred even for our streamlined body discussed in Chapter 7.

Looking at Figure 10.4, it becomes obvious that the strong shock wave and other disturbances must be detrimental to proper circulation (see Chapter 8). In turn, the lift will also be affected near the speed of sound. The much smoother airflow about the supercritical wing of Figure 10.5 is less destructive to lift, and both L and D are improved. With the invention of the supercritical airfoil, economical flight at reduced drag in comparison to the drag on a standard wing was now possible close to the speed of sound (see Figure 10.1).

Many airplanes incorporating these findings have now joined the fleets of the airlines. And although it does not look like it, the flow photograph in Figure 10.4 is in fact that of a supercritical airfoil. The picture was taken at an off-design Mach number, but under design conditions—in this case at a slightly lower Mach number—the same airfoil shows none of the strong shock structures of Figure 10.4. Calculations of the pressure distribution on the wing display remarkable agreement with experimental results, and the Mach number of the experiments in Figure 10.4 is roughly equal to that of current large commercial airplanes. This is another success of modern computational methods, where difficult mixed flow fields are handled. Theoretical fluid dynamics has made substantial progress beyond the earlier results of flow over a wing at low speed shown in Figure 8.14.

The reduction of drag of the supercritical wing, a reduction that depends on the details of the design, applies only to flight at a fixed cruising Mach number. The airliner must, however, fly a part of its total flight time at lower or higher Mach numbers without the benefit of the supercritical wing flow. In Chapter 11.2 we shall discuss how one might be able to get around this restriction in the future, alleviating this and other problems. A practical application of the drag reduction of the supercritical wing applies to compressors whose rotational speeds are sufficiently high to push their blade tips into the transonic regime. Such a compressor can now be designed to operate optimally at a fixed number of revolutions per unit time, and the efficiency of the whole engine is improved.

We shall soon see that large-scale commercial flight at supersonic speeds is unlikely. But moving from M = 0.8 to M = 1 seems at first glance attractive. On long flights, say from New York to Australia, a 15 to 20% speed increase would result in a substantial shortening of the flight time. Aerodynamically speaking, there is no reason why M = 1 airliners cannot be built. However, the substantial increase of the drag coefficient of an airplane in the vicinity of M = 1, the legendary sound barrier, makes flying commercially unreasonable.

We noted that in compressible-flow aerodynamics, model testing must be performed with Reynolds- and Mach-number duplication of the prototype. This is particularly important in the development of further improvements like the ones we have discussed. The exact modeling of the Reynolds number is essential to handle friction coefficients, adaptation of the wing shape to a given flight state, improvement of control surfaces, design of engine inlets, and the like.

Let us consider some numbers that define the modeling problem. For aircraft or birds and bugs (see Figure 6.3), the characteristic length in the Reynolds number, $\bar{c}$ (see Chapter 6 and Appendix 2), is obtained by division of the wing area by the wingspan. On this basis, the Reynolds numbers of large jet airliners range from roughly 10^7 (ten million) to about 10^8, as seen in Figure 6.3. The biggest of them all, the Boeing 747, has a wing area of about 500 m^2 and a wingspan (distance from wing tip to wing tip) of about 60 m, and so we find $\bar{c} = 500/60 = 8.33$ m = 27.3 ft. Flying, for example, at M = 0.8 at an altitude of 12 km (39,000 ft) and making use of Table 4.1 and viscosity data, we find $Re = 5.5 \times 10^7$ or 55 million.

This value of Reynolds number is located on the graph in Figure 6.3 in the region marked "jet transports." It is a fact, however, that models of this and comparable airliners have *never* been tested at the correct high value of Reynolds number at which they cruise. Surely these aircraft, marvels of modern aerodynamics, fly safely. However, the current transonic wind tunnels, while operating at the Mach number of the prototype, impose limitations on model size and operating conditions that make it impossible to achieve perfect duplication of the ratio of inertial to viscous forces expressed by the Reynolds number.

Low-speed wind tunnels such as that shown in Figures 6.1 and 7.11 operate at atmospheric pressure and temperature. The Reynolds number of a given model is changed by a variation of the flow speed in the test section. Even if full-scale Reynolds numbers cannot be achieved, reliable testing is still possible, as testified to by the many flying aircraft. This success is based on an extrapolation of the results obtained at lower Reynolds numbers. But there are limits to the accuracy of the results when it comes to the highly sensitive aerodynamic details listed before.

For wind tunnels operating at higher speeds, including those that produce transonic testing conditions, the pressure in the tunnel circuit can be raised. This increases the density and consequently the Reynolds number. These possible variations in achieving similarity between model and prototype are discussed in Appendix 2.3. Again there is a limit; the wind-tunnel circuit is a large pressure vessel, and to go beyond about five times atmospheric pressure is not practical. In the transonic range, models are supported by rods, or stings, inserted in the back to avoid interference with the flow. At high dynamic pressure (remember q from Chapter 5) and high lift and drag, there is a second structural limit, aside from that of the wind-tunnel walls, to supporting models at high angles of attack.

In sum, the flow velocity is governed by the required Mach number, the density is limited by the pressure, and the model size in relation to the test-section size is constricted by the strength of the model support. The only remaining way to increase the Reynolds number is to *decrease* the viscosity that appears in the denominator of the definition of this parameter. The viscosity of gases behaves in an exactly opposite manner from that of liquids. For example, oils or honey become less viscous when heated, a fact that, as we have seen, becomes important in the choice of motor oil for an automobile in the summer. The viscosity of gases, however, is *lowered* by cooling them, for subtle reasons that can be found in physics books. This fact led to the construction of a *cryogenic* transonic wind tunnel at the NASA Langley Laboratory in Virginia.

This huge wind tunnel operates with nitrogen, which behaves aerodynamically just like air. The nitrogen gas is produced by evaporation of liquid nitrogen at cryogenic (very low) temperatures of about −173°C = −279°F. This temperature lowers the viscosity substantially. The combination of large size, high pressure, and low viscosity permits experiments in the Reynolds-number range required to duplicate the Reynolds numbers of large jet transports, as shown in Figure 6.3. In spite of substantial practical difficulties encountered in the operation of such a large cryogenic transonic wind tunnel, the way is now open for further detailed studies under perfect similarity conditions to bring new aerodynamic advances to flight.

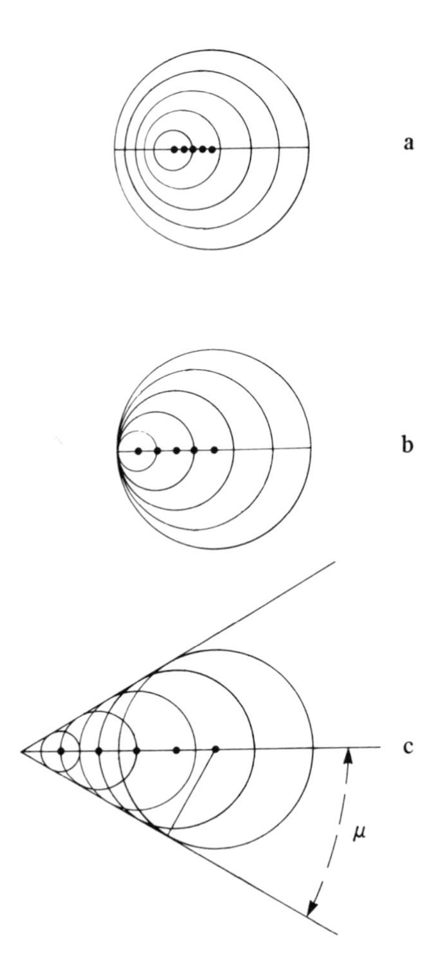

Figure 10.6. *Acoustic patterns of flight. (a) High speed (but subsonic). (b) Sonic speed. (c) Supersonic speed.*

10.2 From Supersonic Transports to Future Dreams

Return to Figure 10.1, where the increase in flight speed abruptly ceases around 1960. From then on, most airliners fly at the same speed in the stratosphere, a speed that is roughly equivalent to M = 0.8. On the other hand, everyone is aware of the supersonic British-French Concorde, which cruises at M = 2 at an altitude of 18 km (about 60,000 ft), substantially above the flight levels of the other commercial aircraft. Calculating the speed of sound at this altitude in the atmosphere (see Table 4.1), we find that for M = 2, the Concorde has a cruising speed of 1,320 mph versus the roughly 530 mph for the others. The supersonic aircraft is therefore about two-and-one-half times as fast! Why don't we have fleets of such airplanes filling the skies?

From the previous remarks on the appearance of shock waves on wings and a sharp increase of drag in the transonic range, we got some feeling for the fact that high speed makes it tough for the aerodynamicist. So far we have talked about flight speeds lower than that of sound. Now we have to understand the drastically different behavior of supersonic flow that we alluded to in Chapter 9.

The three graphs in Figure 10.6 will help us solve our problem. Assume we have a device that produces a series of weak acoustic disturbances, like a hand clapping at regular time intervals. In graphs (a) to (c) we note dots on a straight line, each dot representing the moment that such a disturbance is produced. The motion from right to left of the device creating the disturbances is shown at three different speeds, increasing from top to bottom. Consequently, during the equal time intervals elapsing between the signals, larger distances between the dots appear from top to bottom. Each of the noise signals is transmitted outward at the constant speed of sound. In reality, of course, a continuous noise source such as a moving airplane would transmit its noise continually; we have simply singled out a number of specific moments of this effect.

Return to graph (a). The signal produced at the location indicated by the first dot on the right has after some time created a spherical envelope shown by the outer circle. The signal from each succeeding dot lags behind, but all signals precede the actual locations of the signal generator. Graph (a) depicts motion of the signal generator—the noisemaker—at *subsonic* speed, a speed that is lower than that of sound. But we are not far from this value, as seen by the crowding of the signals advancing to the left.

Moving to graph (b), we note that the distance between the equal time intervals of the signal generator has increased because its speed is higher. All sound signals coalesce on the left. We are moving exactly at *sonic* speed. The coalescence of the sound pulses is a prelude to the shock formation of Figures 10.2 and 10.4. The wave pattern of graph (c), finally, looks radically different. The signal generator outruns its own signals, since it moves faster than the speed of sound. Its motion is called *supersonic.*

At supersonic speed, all sound signals are confined to a conical shroud called a *Mach cone.* The lines bounding the cone are *Mach lines* or *Mach waves.* Viewing Figure 10.6c, we see that two different regimes emerge. Theodore von Kármán, whom I have quoted frequently before, suggested calling the inside of the conical field bounded by the Mach cone the "zone of action." This is the space in which things happen. He next named the outer region the "zone of silence." A noise detector mounted outside the cone does not register that anything is going on until it is passed by the cone. At that point noise instantaneously follows silence.

The angle of the Mach cone is of special interest. Take the first point on the right in graph (c), where a line is drawn at a right angle to the cone. The length of this line is given by the product of the speed of sound, a, and the time interval, t, for the sound to reach the Mach cone. (Speed, L/T, times time, T, gives length, L, from our dimensional reasoning.) During the same time interval that the sound source has outrun its sound, it is now located at the tip of the cone. The distance from the first dot to the tip is given by the product of the speed of the sound generator, V, and the time interval, t. The *Mach angle,* μ (mu), is defined as the angle between the Mach wave and the centerline, as shown in (c). From our triangle we find, using the definition of the Mach number (see Chapter 6),

$$\sin \mu = (at)/(Vt) = a/V = 1/\mathrm{M}.$$

This beautiful equation tells us much about the behavior of supersonic flow. The Mach angle becomes smaller with increasing Mach number. The Mach cone becomes increasingly thin, and the zone of action is restricted to a narrow region. The shock wave's behavior, which we will discuss next, follows the same trend; the tiny meteors or shooting stars we see entering the atmosphere at very high Mach numbers leave a narrow trail.

In Figure 10.6 we deal with acoustic phenomena that are in part well known to us from experience. If high-speed airplanes such as fighter aircraft fly by, we look instinctively in the direction from which the noise is reaching us. But we do not detect the airplane in this direction; it is far ahead by the time its sound has reached us. In addition, the noise of an airplane flying at high subsonic speeds is heard from afar, becomes louder, reaches a maximum, and subsides slowly. These experiences do not apply to an object moving at supersonic speeds.

In Figure 10.7 we have a photograph of a cone-cylinder body flying at M = 3. In contrast to our system of weak waves in Figure

Figure 10.7. Photograph of a cone-cylinder body in flight at M = 3.

10.6, we now see a strong, oblique shock wave starting at the tip of the projectile. This shock wave is an extremely thin front of discontinuous changes of pressure, temperature, and density, all of whose values increase. This jump in density is again made visible by an optical system similar to that discussed for Figures 10.2 and 10.4. Like them, this picture was taken at an exposure time of one microsecond to freeze the rapid action in place. The conical shock caused by the sharp deflection of the flow around the cone now replaces the weak Mach wave of Figure 10.6c, transmitting the pressure jump to locations far away from the flying object.

The photograph reveals weak disturbances after the shock, starting at the ragged edge of the turbulent boundary layer formed on the cone. An expansion follows at the shoulder leading to the conical surface, and a second strong shock wave originates where the flow is turned in the horizontal direction at the edge of the wake. In the wind-tunnel context we say that the parallel flow ahead of the object must become parallel again.

The turbulent wake following the blunt base is remarkably similar to the wake behind a rowboat. Many features of viscous flows, boundary-layer structure, wakes, and the like are remarkably similar in low-speed and supersonic flows. In detail, of course, the effects of compressibility alter the behavior. However, no such dramatic changes appear, such as those shown in Figure 10.6 and described in Chapter 9 for nozzle flows.

At some distance from the body moving at supersonic speed, two strong shock waves remain. This is equally true for a high-flying supersonic aircraft such as the Concorde. The shock waves are transmitted through the atmosphere from great altitudes, and the shock pattern shown in Figure 10.8 emerges. When the shocks graze the surface of the earth, the famous sonic boom appears. Most of us will at some time have heard such a boom, usually produced by a military airplane flying at great altitude. The shocks are refracted in the atmosphere, just like a ray of light passing through a crystal. Depending on the temperature distribution in the atmosphere, the shocks are weakened or strengthened during their passage from the aircraft to the ground.

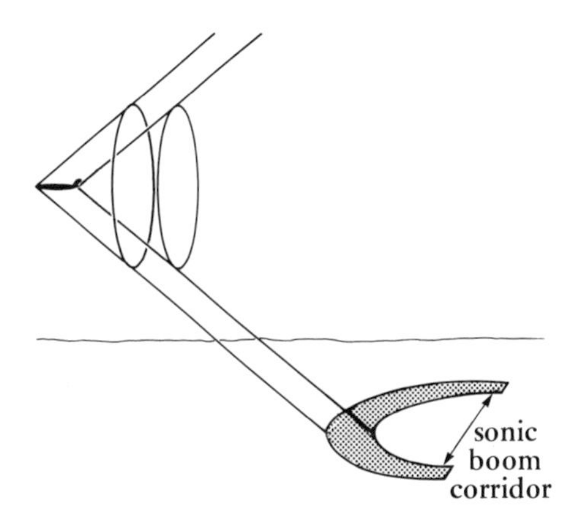

Figure 10.8. Schematic sketch of the origin and effect of a sonic boom. Note the horseshoe pattern of the pressure signal on the ground.

Looking more closely at Figure 10.8, we see a horseshoe-shaped imprint on the ground. This "footprint" represents a cut through the two conical surfaces generated by the two major shock waves. The left boundary of the shaded footprint indicates where the zone of silence is abruptly disrupted by the zone of action passing over us on the ground. An N-shaped pressure pulse of the length of the shaded zone (depending on where we are with respect to the flight path) passes by at the speed of the airplane. The bow shock causes the pressure to increase abruptly, next the pressure decreases below the ambient value, and at the end the pressure returns to its starting point. The footprint passes in about 0.1 second. Unless a large airplane is involved, only a single crack—the sonic boom—is heard. The width of the imprint depends on the altitude of the supersonic aircraft; a system of sonic-boom carpets would be laid down by many aircraft.

The actual pressure signal is not large in absolute terms.* It corresponds roughly to the pressure difference between the top and bottom floors of a tall building (remember how your ears pop in the elevator) or the noise of thunder in the immediate vicinity. Extensive testing series using supersonic military aircraft were carried out by the U.S. government in the 1960s. Overflights at supersonic speed involved two major cities. The effects on the ground were monitored, and it was concluded that sonic booms could not be tolerated. Accordingly, in the early 1970s supersonic flights over inhabited areas were outlawed by Congress.

*It is difficult to give absolute values for the pressure disturbances of an N-pulse of about 0.1-s duration. Many variables enter, such as altitude, flight path (cruising, descending or ascending, turning), weather conditions, and the like, leading to a wide range of sonic booms.

And yet one supersonic commercial aircraft still visits U.S. airports: the Concorde, whose speed is not recorded on our Figure 10.1 because it is not representative of current airliners.

The joint British-French development of this first successful supersonic airliner took many years, in which formidable aerodynamic problems were brilliantly solved. The chosen configuration was the delta wing shown in Figure A2.3, and the aircraft is well known from many illustrations. The delta wing represents a different flow field from that of the swept-back wings discussed before. The new shape presents a host of difficult aerodynamic problems. Large vortices curl up at the edges of the extremely swept-back wings. Separation becomes a dangerous possibility. Based on previous experience with "flying-wing" airplanes, craft without a fuselage, and triangular wings and on much testing in wind tunnels, these problems were overcome. In any case, landing the airplane is carried out with separation in a kitelike position. To retain the view of the runway for the pilots, the nose of the aircraft must be turned down.

Ideally we ought to have a straight wing for takeoff and landing at the lowest possible speeds. For cruising in the supersonic mode, such a straight wing could be folded back to approximate the delta-wing configuration. In fact, military aircraft with variable sweepback exist; the wings are pivoted during flight. Military aircraft also brake their high landing speed on the runway by popping out a parachute in the back. But airliners must abide by different rules, and such schemes are not permitted. There are further problems peculiar to supersonic flight, such as a rapid adjustment of the control system and the trim (see Chapter 8) in the short time elapsing from start to cruise. New developments in the design of powerful engines and the aerodynamics of proper engine inlets in the wide speed range were required. It took some time before the sound level—quite aside from the boom problem—was sufficiently reduced for the Concorde to receive permission to operate at airports in the United States. To study these additional complications, we must refer to the special literature on this interesting aeronautical development.*

The currently existing Concordes in the British and French fleets are not being added to with new models, and we shall see in the next chapter what the future might bring. The former Soviet Union, with its vast empty spaces and long distances to be covered by aircraft, supported a similar development of a supersonic airliner. Designed by the Russian Tupolev, who had contributed much to Soviet aviation, the TU-144 shown in Figure 10.9 looks remarkably like the

*For a fascinating account of how to fly a Concorde across the Atlantic, in which all the special problems that arise are discussed, see Richard L. Collins in the magazine *Flying,* April 1985.

Figure 10.9. *The Tupolev TU-144 supersonic transport on the runway.*

Concorde. Although the Russians must have learned much from the published research of the West, their solutions to aerodynamic problems dominating the outer shell of the aircraft were driven by engineering necessities. One TU-144 crashed, and at this time the airliner does not operate. The supersonic projects of the United States contemporaneous with the development of the Concorde were more ambitious: a much larger airplane at M = 3 was planned. This work was discontinued when the practical problems discussed above appeared.

A last remark on the special phenomena of supersonic flight pertains to the problem of aerodynamic heating. Thermodynamics enters aeronautics when aircraft move ahead in speed into the compressible range of fluid dynamics. The air brought to a halt at the nose and on the surface of an airplane is heated by compression. That effect is negligible at low speed. Even for an airliner in the transonic range flying at M = 0.8 in the stratosphere at the ambient temperature of –56°C (–67°F) found in Table 4.1, the ram temperature is computed to be only somewhat higher at –28°C (–18°F). Such calculations arise from a solution of the equations of motion of gasdynamics that we mentioned but cannot reproduce here. Touch the skin of your aircraft after landing and you will find it quite chilly.

At M = 2, however, the Concorde flying in the stratosphere is heated to 118°C (244°F), a temperature *above* the boiling point of water. This heat is transferred by an ingenious cooling system to the rapidly consumed fuel of the craft. The effect of aerodynamic heating lengthens the Concorde through expansion of the skin by about one foot in supersonic flight! The surface temperature approaches the upper limit at which aluminum alloys remain sufficiently strong to be structurally sound. The U.S. supersonic transport would have developed a surface temperature of about 335°C (635°F) at M = 3, a temperature that dictates the use of titanium-based alloys for those parts of the airplane that are exposed to the heat. Such metals retain their strength at elevated flight temperatures, while aluminum reaches its melting point at 660°C (1,220°F). We can see that new shapes and the heat problem dominate aerodynamics in the supersonic regime.

This brings us to our last remarks in this chapter, which deal with *hypersonic* flight, the frontier of aerodynamics at extreme speeds. "Hyper-" is derived from the Greek, while "super-" comes from the Latin. Actually both prefixes have the same meaning; they refer to flight above the speed of sound. However, the appellation *hypersonic* has traditionally been used to describe the flight regime above a Mach number of about five. Several coincidental aerodynamic effects appearing around M = 5 set hypersonic flow apart from the well-understood supersonic range. For example, the temperature in the boundary layer and on the surface of flying objects increases steeply beyond that on the supersonic aircraft we have considered. At M = 5 the ram temperature on the nose of an airplane flying in the stratosphere rises to 1,000°C (over 1,800°F).

Under such conditions the heated air ceases to be the well-behaved, inert gas mixture that we described in Chapter 4. The atoms in the molecules of oxygen and nitrogen begin to vibrate, jiggling against each other. This internal molecular motion leads to serious aberrations in the thermodynamic behavior of the two gases. Further increases in Mach number heat the air adjacent to the aircraft to such high temperatures that oxygen and nitrogen *dissociate;* the heat destroys the molecular bond and frees the atoms to go their own ways. With oxygen and nitrogen atoms present in the flow, new compounds are formed by chemical reactions, further complicating the behavior of the medium we fly in. The *simple* gas laws that relate the variables of pressure, density, and temperature to each other become invalid. The aerodynamicist who designs a space shuttle or hypersonic intercontinental missiles must turn into a physical chemist and consider previously unknown territory of the fluid dynamics at extreme temperatures.

But that is not all. The highly narrowed zone of action at such speeds—consider the Mach angle at M = 5 and higher—causes boundary layers and shock waves to be jammed together. The notions developed in Chapter 5 suggested a way to deal with regions of inviscid flow bordered by a narrow boundary layer in which viscosity cannot be neglected. These descriptions of the flow field fall by the wayside. Turbulence and viscous effects tightly shroud the body, the air turns into a fiery mess of reacting gases, and extreme rates of heat transfer to the surface of the flying machine must be coped with.

Because of all that, model testing becomes very difficult. Even if we separate the different effects appearing in this regime and plan to study the aerodynamics at high Mach numbers first without the real heat effects, we encounter problems. In hypersonic wind tunnels, the speedup to these Mach numbers is associated with a drop in temperature in the test section that makes the air condense, the very opposite of the effect of free flight. Thus the air for the wind tunnel has to be preheated. To study the full range of effects, however, we need to duplicate the high temperatures in the flow field in addition to Mach and Reynolds numbers. This feat cannot be accomplished with current technology, and is not yet within our grasp. There has been some success, however, in modeling Mach number and temperature (not Reynolds number) simultaneously in a special apparatus in which the testing time available is reduced to fractions of seconds.

In practice, aerodynamics at hypersonic speed is first studied in a hypersonic tunnel that does not model the high temperatures. With these results, preliminary designs are checked out in a high-temperature facility. All experiments are guided, of course, by theory and computations for given prototypes to ensure a functioning device. Moreover, the aerodynamic aspects—as always—do not stand by themselves. The design of propulsion systems and structures for hypersonic airplanes and missiles requires wholly new approaches. The required thrust to get to high speeds and methods to keep the craft from burning up must be mastered. The hypersonic range has opened up a new, fascinating chapter of aerodynamics that mimics the flight of meteors.

After all this has been said, it is remarkable that hypersonic reentry of the atmosphere by space capsules flying on ballistic trajectories has long been successful. We even have one hypersonic, controllable manned vehicle. The space shuttle is shown in a friendly environment in Figure 10.10. This photograph is chosen to demonstrate the large size of the shuttle, which with a length of 37 m (122 ft) and a wing-

Figure 10.10. *Return of the STS-2 orbiter* Columbia *to the Kennedy Space Flight Center aboard the Boeing 747 shuttle-carrier aircraft.*

span of 24 m (78 ft) is longer than a Boeing B-737 airliner and nearly reaches the 737's wingspan. In the photograph, *Columbia* rides piggyback on a special version of the currently largest airliner, the B-747. Note the additions of large vertical endplates to the horizontal stabilizers (see Figure A2.1) of the carrier plane. The turning of these two aircraft into one flying machine is an amazing aerodynamic feat.

The space shuttle lifts off pushed by a rocket engine that operates with a liquid propellant, just like the rocket shown in Figure 9.4. The fuel and oxidizer are carried in two tanks enclosed in a huge container. The orbiter itself, the aircraft that returns to earth and is shown on top of the carrier vehicle in Figure 10.10, is attached to the big container. Liquid hydrogen is the fuel and liquid oxygen the oxidizer. That combination is the most efficient propellant available. It is the external container that dominates the assembly when you view the television pictures of a space-shuttle liftoff. To further assist the start, two large solid-propellant (think of gunpowder) booster rockets are attached to the sides of the large container. Each of these generates an additional thrust (see Chapter 9.2) of nearly three million pounds (13×10^6 N; see Table A3.3).

The two booster rockets are dropped off at an altitude of over 40 km (25 mi), to be recovered for another flight. The big container is separated later, shortly before the shuttle reaches the *orbital* speed of about 8 km/s (5 mi/s = 18,000 mph),* and it burns up during its fall to earth. The entire system is designed to put the space shuttle into previously selected orbits at altitudes up to 1,200 km (750 mi) or so. At such altitudes there is practically no air, and thus the aerodynamics we have looked at does not apply.

It is on the return trip that things become interesting from our viewpoint. The shuttle encounters increasing density up to sea-level values (see Table 4.1). Somewhere along the reentry trajectory, hypersonic atmospheric flight with all its attendant problems is encountered. Two orbital-maneuvering rocket engines fed by propellants stored on board and producing 6,000 lbs (27,000 N) of thrust can be used for short periods of time. This push represents only about 1/500 of the thrust of one of the two booster rockets. No new flight plans can be pursued; sufficient power to ensure a safe return is the sole object. In addition, tiny rockets can be fired in different directions to adjust the flight path before control surfaces such as those shown in Figure A2.1 come into play in the dense portion of the atmosphere.

---

*Spacecraft that leave the earth rather than orbit it require an *escape* speed of over 11 km/s (6.8 mi/s = 24,600 mph) to overcome the gravitational pull of our planet.

Prior to reentry, the craft is slowed to depart from its orbit. It is then pointed toward a carefully planned reentry corridor aimed at an exact landing site. The corridor looks like a funnel, with a large opening at the top and a tiny spout located at the point of touchdown on the runway. If the flight path becomes too steep, the orbiter will burn up like a meteor. Should the orbiter overshoot the reentry funnel, it will not be able to hit the landing site. In the latter instance, its further entry maneuvers will be dictated solely by the laws of physics, since at that stage of flight the space shuttle is an enormous sailplane without primary propulsion.

The sequence in a normal reentry goes as follows. At first the craft dips down and encounters increasingly dense air at Mach numbers well above ten. Soon temperatures higher than the surface temperature of the sun (6,000°C or 11,000°F) appear at the front of the shuttle and in the boundary layer. Extreme rates of heat transfer set in. However, such rates need to be handled for only a short period of time. This is in contrast to the heat—albeit at much lower values—that is encountered on a continuous basis by the Concorde. The spacecraft is insulated by the famous ceramic tiles; this insulation keeps the inside cool for a short period of flight at these extreme conditions.

At the high point of the reentry sequence, the kinetic energy of motion converted to heat is sufficiently high to alter even the atomic structure of the gases surrounding the shuttle. The electrons are knocked out of their orbits around the nuclei. A brightly glowing plasma shrouds the space shuttle, and sharp detonations are heard by the crew whenever the tiny course-control rockets are fired; one astronaut likened this phase of flight to a dive into Dante's inferno. Viewed from the ground, the trajectory is lit up like that of a meteorite. At this stage no radio communication is possible: the plasma is impenetrable by radio waves.

Finally, the dense air produces aerodynamic drag to slow down the large aircraft. Skilled piloting steers it to the runway for a conventional touchdown. In sum, hypersonic flight with all its problems needs to be braved only during a small fraction of the time of a space shuttle's mission.

Returning to the earth, we recall that current airliners—excepting the handful of Concordes—fly close to, but below, the speed of sound. Further advance of flight speeds has been blocked by the sonic-boom problem. Although much research to find ways for commercial airliners to avoid creating booms in atmospheric flight has been carried out in many countries, no remedy has been found, and, as stated before, none is in sight. Now if we could rise above the clouds and the dense layers of the atmosphere, fly at extreme altitudes, like an orbiter, and reenter the atmosphere to land at a distant airport, the boom problem would vanish. Indeed, an aerospace airplane that might do just that has long been discussed in aeronautical circles, and various designs have been considered in the United States and Europe. But leaving out these far-fetched dreams, we must not overlook the enormous progress in earth-bound and space flight in a remarkably short time. Supersonic speeds are now routinely achieved by the Concorde, and hypersonic flow is mastered by the space shuttle.

There is a certain magic in the span of two hours in relation to our endurance and enjoyment of travel. The idea of reaching any point on earth from any other point in two hours has fascinated aeronautical engineering. The often-invoked criterion of speed by itself has serious limitations. Speed is not felt by people. We turn on the earth once in 24 hours, move around the sun on a huge orbit once a year, and fly with the sun's orbit about the center of our galaxy at a speed of 220 km/s (140 mi/s = 5×10^5 mph!) without noticing anything. Newton's laws tell us that it is acceleration—not speed—that accounts for our physiological responses. Therefore, not speed but time and comfort govern our endurance and enjoyment of travel.

The aerodynamicist and airplane designer Küchemann, whom we quoted in the epigraph for Chapter 8, studied the frequency of marriages in Oxfordshire (England) as a function of distance and travel time between villages.* The maximum distance over which marriage partners were found was remarkably constant over the two hundred years between 1650 and 1850. It remained at 10 km (6.2 mi), or a travel time of about two hours. With the advent of the railroad, the distance increased, yet the time of travel remained at two hours.

Taking two hours as a limit, commercial air transportation of the kind envisioned for the aerospace plane would fill the bill: no boom problems and two-hour flights between any two places on earth. A backlog of existing technology based on military airplanes, missiles, and space flight is available to get started on the aerospace airplane. Is there anything involved that we do not know? Unfortunately, the answer to this question points to many unknowns. The aerodynamics of sustained high-altitude flight at such speeds—our own interest—may well turn out to be the least of them.

Aside from the fact that the plans for the aerospace plane are more complex than those for any previously developed airplane, the craft has to perform the extra function of flying into orbit under its own power *without* external tanks and boosters that can be jettisoned. Propulsion will be a primary problem. The liftoff weight of the complete space-shuttle system is 4.4 million pounds, while that of a fully loaded B-747 is about 800,000 lbs. About 3.7 million pounds—that is, over 80% of the total weight at the start—is fuel and oxidizer stored in liquid form at extremely low temperatures. The oxygen ac-

*See the book by Küchemann cited in Appendix 4.

counts for over 80% of the propellant weight. The novelty of the aerospace plane is that it would carry mostly fuel, relying primarily on the oxygen in the air for combustion. Only at high altitudes above 60 km (40 mi) or so, where hardly any air is left, would rocket engines give it a final push to attain orbit if needed for a given flight.

Thus air-breathing engines such as those discussed in Section 9.2 reappear. We mentioned the efficient ramjet that operates at supersonic speed. The aerospace plane will have to be pushed off the runway by traditional jet turbines to the speed where ramjets take over. The aerodynamics of the airplane will be much affected by large air inlets that devour the air needed to oxidize the ramjet fuel at supersonic speed. A final extension of the ramjet principle is further contemplated. In the conventional ramjet—as in the jet turbines for supersonic aircraft—the ram air is slowed to subsonic speed by a diffuser. The fuel is added and burned in a combustion chamber much like that of a jet turbine. The new engine must be designed to receive and burn fuel in a supersonic airstream. Think of the problem of lighting a match in a storm. Devices that can do this trick, called "scramjets," must be developed beyond currently tested small model versions. To design a combination of practically all modes of propulsion to power a single airplane will require the combined efforts of several nations.

Clearly, all this is still far in the future. Military interests enter, just as they do for the space shuttle, and development cost is highly uncertain; these are matters that are far off our topic. The extension of our knowledge of low-speed aerodynamics to that of the compressible, subsonic flow of current airliners has left us with our basic ideas intact. The extension of the speed range to supersonic or even hypersonic flow could be discussed only qualitatively, but since we are now aerodynamicists, it was important to understand a little of flight in these regimes. Newspapers, magazines, and television bombard us with futuristic schemes. These notes may help us to adopt a more skeptical view than that we encounter in the popular press. More about this in the final section of the last chapter.

Chapter Eleven

Air Transportation and the Outlook for the Future

I shall not attempt to argue with those who consider any increase in speed unnecessary. The public will always prefer that conveyance which is the most perfect, and speed within reasonable limits is a material ingredient in perfection in traveling.

Isambard Kingdom Brunel (1806–59)

11.1 The New Traffic Patterns

In this section we shall leave aerodynamics for a while to turn to aviation. Let us look briefly at domestic and international passenger flight. There exists an extensive, readily accessible literature on the subject, and so we will view only a few interesting—and lesser-known—facets of the traffic revolution occasioned by aviation. Airplane buffs who are interested in the great variety of existing and planned aircraft for civilian and military use should consult the works listed in Appendix 4.

First let us study aircraft in the context of other modes of locomotion viewed strictly from an engineering viewpoint. In 1950 Gabrielli and von Kármán published a study of what they called "the irresistible urge of humanity toward increasing the speed of locomotion. Means of locomotion on the ground, on the surface of and within

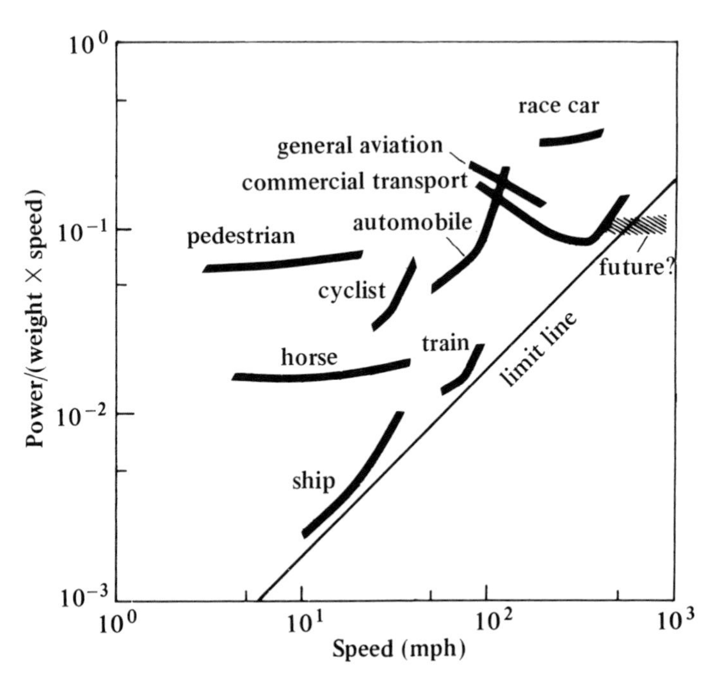

Figure 11.1. The transport economy index (see text) as a function of speed for different modes of locomotion.

water, through the air, and perhaps through empty space, compete in an ever-growing effort toward higher velocities."*

The basic premise of their remarks has not changed in the intervening years. The authors offered a dimensionless coefficient given by the power needed to propel a vehicle divided by the product of its weight and maximum speed. Using Table A2.2, we readily confirm that this coefficient is indeed dimensionless. It has also been called a transport economy index because it compares the efficiency of types of transportation in relation to their purpose. Selected and extended values of this index as a function of speed are shown in Figure 11.1.

The results of this study need to be viewed with care. For example, not all types of transportation must be fast. Ships are slow, but nothing beats them for carrying heavy loads from here to there. A horse does better than a man at carrying loads, a fact that has been obvious throughout history. On the other hand, race cars do not need to be efficient, and shipping costly orchids is best done at the highest speed possible without much concern for efficiency. Speed is tied to

*G. Gabrielli and T. von Kármán, "What Price Speed?" *Mechanical Engineering* 72:775 (1950).

tractive power per unit weight, and a price must be paid to increase the speed, which may or may not be important.*

Aircraft show a decreasing economy index in a certain speed range. The index picks up steeply again toward the current airliner flight speeds of around $M = 0.8$. An empirical limit line drawn at 45° in our graph is suggested by the values for the moving objects shown. Beyond this line, a further speed increase might render a specific vehicle impractical. Küchemann (see citation in Appendix 4) theorized that supersonic and hypersonic airplanes of the type discussed in Chapter 10.2 might break through this barrier in the future. As shown, their economy index would be somewhat above that of a pedestrian!

For a comparison of modes of transportation, a complex interplay of speed, the elapsed time of a voyage, the distance covered, the cost of travel, personal convenience, and so forth must be considered. This task transcends our discussion of the engineering aspects of the transport economy index, but it is a fact that for many years, airplanes have prevailed in transporting people, mail, and a variety of goods. In the period 1926–86, over 150 different types of aircraft (including revised versions of existing airplanes) were flown by U.S. airlines. In the same period the surprising number of about 150 airlines were founded in the United States. Of these, over 70 were in business in January 1986.† This number includes the large companies whose airplanes connect major cities in the United States and most countries on the globe in competition with the fleets of other countries, as well as commuter airlines connecting smaller cities to each other and to the big hubs. In addition, there are local airlines that have just a few aircraft to provide sightseeing flights and the like.

In the 1990s, there have been many changes in the pattern of commercial aviation. Here and abroad we find new airlines being founded while some other airlines have been involved in mergers and bankruptcies. To better adapt airplanes to desired purposes, closer relationships between airlines and aircraft manufacturers have evolved. Even before ordering an aircraft, airlines take part in the process of determining its technical specifications, not to mention the design of the interior of the jetliner, from the color of the cabin to the location of the microwave ovens and the leg room between the rows of seats.

---

*Speed appears in the denominator of the efficiency index. One would think offhand that high speed therefore decreases the index. However, we recall from the previous chapters that power increases much more than linearly with speed. Therefore the index increases steeply with increasing speed.

†Myron J. Smith, Jr., *Passenger Airliners of the United States* (see citation in Appendix 4).

New carriers now serve the smaller cities abandoned by the major airlines. Such service is provided by smaller airplanes, which carry from about fifteen to over fifty passengers and are mostly powered by turboprop engines. It is interesting to note that many of these short-haul aircraft are manufactured abroad in countries as far apart as Ireland and Brazil. Flying the large airplanes, in turn, we now often have to change aircraft at a hub rather than flying nonstop. Usually this routing increases the time of travel. Airlines whose home base is in the United States increasingly have cooperative agreements with European carriers to provide them with rapid access to cities in the interior of the United States. In the People's Republic of China, the national airline has been broken up into different regional carriers, and the giant former Soviet air fleet struggles with reorganization and modernization. This complex situation is in flux, and we shall not attempt to foresee how such organizational problems will be resolved.

The development of this vast aviation system has been based in large part on the progress indicated by the rapid increase in the performance of large passenger jetliners, some of whose characteristics are shown in Table 11.1. Here we concentrate on selected airliners; however, some small planes and their forerunners the birds are included. Comparing the Wright *Flyer I* of 1903, the first flying machine that could rise into the air under its own power with a pilot, to the Boeing 747 of 1971, the largest passenger airplane to date, it is remarkable that a period of less than seventy years has elapsed between the flights of these two airplanes, a fact that is typical for the breathtaking rapidity of the advance of aviation. Advances in aerodynamics have played a major part in this development, together with progress in structures, propulsion, avionics, and so forth. All this has led to the current high values of payload and speed. The numbers on wing loading—the weight per unit area that can be sustained by the wing—tell this story. Every square foot (about 0.1 m^2) of the wing of a Boeing 747 can support about forty bricks, or more than three loaded suitcases!

We further note that the cruising speed of airliners introduced since the beginning of the jet age around 1960 is quite uniform. Cruising in the lower stratosphere (see Figure 4.2) at temperatures of −67°F (−56°C), the aircraft fly in the narrow Mach-number range of about 0.80 to 0.85. This range includes the Airbus 320, called the first "fly-by-wire" or "electric" airliner.* In such an airliner the pilot commands a computer system that in fact flies the airplane. This computer system and its redundant backups go far beyond the so-called automatic pilot, a device that keeps an aircraft on a preset course.

*M. Mitchell Waldrop, "Flying the Electric Skies," *Science* 244:1532 (1989).

TABLE 11.1. SELECTED EXAMPLES OF AIRCRAFT DEVELOPMENT FROM THE WRIGHT *FLYER* TO OUR TIME.

	Year	Takeoff Weight (lb)	Speed (mph)	Lift/ Drag	Wing Loading (lb/ft²)
pigeon		0.9	28	5.4	1.3
wandering albatross		21	45	20	3.1
VJ-24 hang glider		310	20	9	1.9
Nimbus-3 sailplane		1,500	50	60(!)	8.6
Piper Super Club		1,750	72	10	9.8
Cessna 310 F		4,800	125	12	28
Wright *Flyer I*	1903	750	35	8.5	1.5
Ford Trimotor	1927	11,000	110	12(?)	14
Douglas DC-3	1935	25,000	180	15	25
Douglas DC-6	1947	105,000	315	18.5	72
Boeing 707-320B	1959	333,600	530	18	115
Douglas DC-9-30	1966	121,000	525	16.5	121
Boeing 737-200	1969	115,500	465	16	115
Boeing 747-200	1971	833,000	555	18	150
Douglas DC-10-30	1971	565,000	564	18	156
Airbus 300	1974	302,000	530	15	108
Boeing 767-200	1982	315,000	530	18	105
Boeing 757-200	1983	220,000	530	18	115
Airbus 300-600	1984	376,000	540	17	134
Boeing 737-300	1984	124,500	485	16	125
Boeing 767-300	1985	412,000	530	18	135
Airbus 320	1988	162,000	527	17	123
Boeing 747-400	1989	870,000	555	18	155
Boeing 777-200	1995	535,000	550	18	115

The new system is fully automated; computers decide all aspects of flying the aircraft from takeoff to landing by commanding actuation of the control surfaces, monitoring the engines, and so forth. While such systems may well dominate the future, there is still some controversy about the total automation of flying.

In addition to the airplanes in Table 11.1, there are the new Airbus 340, a large aircraft that not long ago entered transatlantic service, and the MD-11 of the McDonnell-Douglas Corporation, an updated version of the well-known DC-10. The Boeing 777, the MD-11, and the Airbus 340 have two, three, and four engines, respectively. The three aircraft were designed to fill roughly the same niche in the spectrum of large airliners. Why do they differ in the number of their jet turbines? Clearly, having only two engines minimizes initial cost, fuel consumption, and maintenance cost. All airplanes must obtain certification by the FAA, the Federal Aviation

Administration, whose word is law for aircraft that wish to operate in the United States. A twin-engine aircraft must meet the special standards of ETOPS (extended twin-engine operation): that is, it must be able to stay in the air with one engine for a recently increased time of 180 minutes. On the heavily traveled North Atlantic route, that time period is necessary for reaching airports that are open in winter in Greenland, Iceland, Ireland, and Newfoundland. At the time of the design of the new Airbus 340, the more powerful engines that push the Boeing 777 to meet ETOPS were not yet available. This story is one illustration of the tough competition among the few manufacturers of large commercial airliners. Such advances would not be possible without the remarkable reliability of modern turbine engines.

A final note on the Boeing aircraft in Table 11.1 relates to the design philosophy of this company. In the planning of a new aircraft, thought is given to its future evolution. Quoting Bengelink (see Preface), by "sizing" the wing in the beginning of the design to be readily adaptable to such evolution, families of airplanes can emerge. In the table we note the Boeing 767-200 and its later 300 version, and the ubiquitous Boeing 737 can be found around the world in five versions adapted to different uses. A wing loading of about 150 lb/ft^2 is viewed as the current upper limit dictated by the available materials. With these scanty notes on modern commercial airliners, let us turn to their use.

The dramatic improvements in airliners have brought about revolutionary changes in public transportation. In 1957 about 20 billion passenger miles were flown by airlines in the United States. This number had increased tenfold by the year 1980. (A passenger-mile is a mile flown by one passenger.) The year 1957 has a special significance. After a peak during World War II, railroad traffic in the United States dropped by 80% to about 20 billion passenger-miles in 1957. This was the year in which rail, bus, and air traffic had equal volumes. Railroad and bus traffic have since dropped even more, and flying—in terms of passenger-miles—has far outdistanced all other modes of public transportation.

International air traffic also rose markedly. In particular, with the start of the jet age around 1960, transatlantic flight forced the large ocean liners out of business. (Obviously, ships are more popular than ever in the cruising business. Occasionally, when a ship such as the *QE2* changes its operations from the Caribbean to the Mediterranean, it still carries passengers across the Atlantic.) A sizable increase in total air transportation is seen when the airlines of other countries are added; indeed, the fraction contributed by U.S. airlines to the total number of passenger-miles flown declined slightly in the period 1970–80.

At the same time, a high fraction of the planes providing air service in *all* countries—excepting those of the former Soviet Union—are manufactured in the United States. Major competition currently arises from Airbus Industries, a European consortium that, as we have seen, produces advanced airliners. So far, however, the United States holds a monopoly on the production of the largest airliners, as represented by the Boeing 747. This craft in all its existing and planned versions remains the biggest of them all, with no competitors in sight. In addition, the new Boeing 777-200 promises to dominate the field in its size.

The statistics on worldwide air transportation, as well as extrapolations to the year 2,000, are given in Figure 11.2. We note that a period of exponential growth (Appendix 1) ceases in the 1980s. To better understand Figure 11.2, additional facts are needed. The total number of airliners has increased somewhat during the period of dominance by air traffic. However, this increase is small in relation to the fifty-fold growth in passenger-kilometers. It is primarily the size of the aircraft that has greatly increased. In 1936, a DC-3 produced 7,350 passenger-km/h carrying 21 passengers at 350 km/h. In contrast, a B-747, carrying 500 passengers at about 900 km/h, achieves 450,000 passenger-km/h. Looking at Figure 11.2, we note that the trend of the exponential increase is broken around 1977. The extrapolated curve beyond the available data ceases to increase as before; its gradient is lower. The curve flattens near the year 2000, and it is predicted that *saturation*—the termination of further growth—will occur sometime early in the next century. A warning about the validity of extrapolations of this sort is of course in order; in aviation many surprising things that are not foreseeable may emerge, as we experienced in the past.

An addition of about 40% of the current values of passenger-kilometers is all that is needed, according to statistical analysis, to achieve saturation, or a stagnant value of air travel. This trend reminds us of the escalation of speed of airliners terminated abruptly by a constant value, as shown in Figure 10.1. If the predicted saturation of the air-traffic market does occur, it will have far-reaching effects on the logistics of flying, airport congestion, scheduling, number of takeoffs and landings, and so forth. If the trend to larger aircraft persists, we may even have fewer airplanes in circulation and less traffic on the runways. The new technology required to handle larger numbers of people at airports at a given time is outside our field. Again—as before—we must be careful to accept such extrapolations in view of possible unforeseen events. Some notes on this and related trends must, however, be added later in this chapter.

So far we have simply listed trends and predictions concerning air traffic. The early history of this development is complex and in-

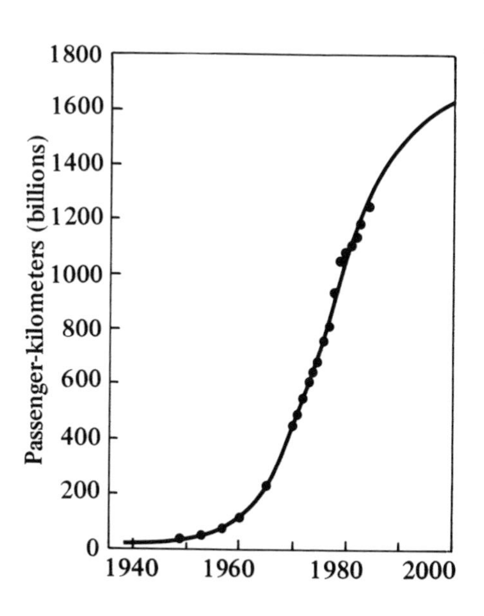

Figure 11.2. *Passenger-kilometers per year of all world airlines, including an extrapolation to the year 2000.*

deed beyond our concern, but a few remarks are in order. Smith* suggests "three distinct chronological frameworks" in which to view the evolution of aviation in the United States before World War II. In the period 1919–26, discussions about how to organize the enterprise prevailed. From 1926 to 1933, the government invested about $500 million (over $9 billion in current terms) in many aspects of aviation. Finally, 1933–40 saw the successful outcome of this generous support. Smith points out that the so-called Lindbergh boom was not the cause of the substantial federal commitments. In fact, Lindbergh's epochal 1927 flight was preceded by major investments such as the government funding of NACA and the private contributions of the Guggenheims (see Chapter 2.1).

These developments culminated in the previously mentioned transoceanic mail and passenger services inaugurated prior to World War II. The war put a stop to expanded efforts to cover the globe with commercial airline routes. The prewar flight services primarily covered the Pacific, with routes like San Francisco to Hong Kong via Manila. But soon after the end of the war, regularly scheduled flights across the Atlantic were established. The long distances to be covered required intermediate stops at locations such as Dublin; Prestwick, Scotland; or Gander, Newfoundland. Nonstop Atlantic crossings are now provided by airlines of many countries, leading to tough competition for what is still the busiest international air route.

The development of airliners does not stand still. Other than aerodynamics, jet turbines, and electronic control systems to handle the actual flying of the airplane, many other aspects enter the whole enterprise. Instrument flying—flying without sight of the ground—was already in use in 1928. Such instrumentation, on the airplane and on the ground, reduces the dependency on the vagaries of the weather.

Of course, even now weather plays an important role in aviation. Weather still affects the intervals at which aircraft can start or land, and airports may occasionally be closed to all traffic. The choice of the most favorable route and altitude of cruising flight also depends on meteorological factors such as the jet streams. Here the avoidance of powerful head winds—or gaining their assistance if they blow in the direction one wants to fly—is important. Further research is required to develop new techniques for detection of violent behavior of the weather. For one, there is the strange phenomenon of clear-air turbulence (CAT), strong gusts

---

*Richard K. Smith, "The Intercontinental Airliner and the Essence of Airplane Performance," *Technology and Culture* 24:428 (1983), and 26:870 (1985).

that are invisible to the eye because they are not marked by cloud motion. Similarly, unexpected strong downward drafts called wind shear occasionally arise suddenly. These are of particular danger during a landing in bad weather. Finally, formation of ice on the wings of aircraft in flight remains a problem, especially for smaller airplanes. However, it is remarkable that the pilots, together with the assistance of air traffic controllers and supported by instruments on the ground and in the air, can handle the dark of night, make practically blind landings at unseen airports, avoid other aircraft, and fly safely through most storms.

Returning to commercial flight, the gap needs to be filled in our brief history between the Wrights and the most recent developments. In a discussion of this sort, the Douglas DC-3 airliner shown in Figure 11.3 invariably appears. There is still some disagreement among historians about the actual importance of this famous airplane. It seems certain, however, that the DC-3 is the first "modern" passenger transport.

One important structural development predates the DC-3. During World War I, the German aircraft designer and builder Hugo Junkers devised a metal wing that could be fastened directly to the fuselage by an internal structure. This wing did not require external bracing by wires, pylons, and struts, and thus it produced a substantial lowering of the aerodynamic drag. Junkers later used corrugated steel sheets for both wing surfaces to provide additional strength. The Junkers JU-52 shown in Figure 11.4, a low-wing trimotor airplane, became the mainstay of the German and other airlines in the early 1930s. (About five thousand JU-52s were built.) The high-wing Ford Trimotor introduced in 1925 had a smooth skin and used a similar method of construction, but both aircraft still had fixed undercarriages. Soon afterward smooth outer skin appeared universally, to be followed by retractable landing gear and many other advances.

But back to the DC-3. Much knowledge of aerodynamics and structures had accumulated during the 1920s, and the airline industry was demanding improved planes. Specifications were drawn up by one airline to cover the required number of passengers, takeoff speed and runway length, amenities, and so forth.* Donald Douglas, who in 1920 had founded a small aircraft manufacturing company that built military airplanes but had never produced a large transport plane like those of the Boeing Company, entered the bidding. The DC series resulted. When the DC-3 entered scheduled service in June 1936, it was the biggest aircraft of its time, much like the Boeing 747 today.

---

*Richard F. Snow, "The Letter That Changed the Way We Fly," *American Heritage of Invention and Technology* 4:6 (1988).

Figure 11.3. A Douglas DC-3 passenger airliner (1936).

Figure 11.4. A Junkers JU-52 passenger airliner (early 1930s).

The Cal Tech wind tunnel shown in Figures 6.1 and 7.11 was used to provide systematic studies of the DC-3's aerodynamics. A retractable hydraulic landing gear, the use of a strong aluminum alloy for the skin, cabin soundproofing, and many other new features characterized this airliner. Two 1,000-HP (746-kW) Wright engines provided propulsion for carrying twenty-one passengers in comfort, with service provided by a steward.

By 1939 the DC-3, of which 10,629 were built, was carrying three-quarters of all domestic airline passengers, and during World War II it played a major role as the U.S. C-47 transport and its British version, the Dakota. It was used in a remarkable range of applications, from transporting paratroopers to carrying freight. DC-3s are still in service today in many parts of the world. However, after the war many new airliners were designed with important improvements such as cabin pressurization, and of course we have been living for about forty years in the jet age.

We shall now turn to a brief view of the future, with aerodynamics again being our overriding interest.

11.2 What Will the Future Bring?

From the Wright *Flyer* to the Concorde, with flights to the moon—including two passages through the earth's atmosphere—and the space shuttle mastering hypersonic, controlled atmospheric reentry, aeronautics and astronautics have made tremendous progress in one lifetime. Based on the experience of such dramatic progress, it is daring to speculate on the future of commercial flying, and yet predictions made now may be safer than those of the past—recall Cayley's thoughts on the potential speed of aircraft—since at this time aviation is based in large part on mature technologies.

Looking at a period extending into the next century, it appears that much of the individual heroics of the past will not recur. As previously mentioned in Chapter 10.2, serious thought had been given to the development of a fleet of supersonic airliners flying at a Mach number of up to three. A demand for 250 to 300 such aircraft was foreseen for the year 2010.* These airplanes would fly long-distance routes, in particular to the increasingly important Far East. Ignoring the daunting cost of such a project, and daringly assuming that an aerodynamic solution could be found for the sonic-boom problem discussed in the last chapter, there still remains a major impediment in the way of the supersonic airliner. Recently our understanding of

*James P. Loomis, director, Center for High-Speed Commercial Flight, Battelle Memorial Institute, Columbus, Ohio, in the *New York Times,* July 10, 1988.

the chemistry of the stratosphere has much advanced. A fleet of larger and faster supersonic aircraft would fly at even greater altitude in the stratosphere than the Concorde (see Figure 4.1). It is now understood that the exhaust products of the powerful jet engines deposited in the upper stratosphere would cause a serious deterioration of this layer, leading to disastrous environmental consequences.

As the next step in line, the hypersonic aerospace plane described in Chapter 10.2 may remain a subject of study. As in the case of the initial research performed for the space shuttle, military interests may enter. However, any realistic thought of seeing us fly in this wondrous machine, for which the technology is *not* in hand, must at best be delayed far into the next century. A factor contributing to this situation, as we shall see, is that much of the near future of commercial flight will be governed by political and social considerations rather than by purely technological innovations.

The development costs of supersonic and hypersonic airliners compete at the federal level with those of military airplanes, refit of existing intercontinental ballistic missiles, satellite technology, and the plans of NASA regarding space science in general, unmanned space probes, moon landings, an orbiting space station, a manned colony on the moon, manned flight to Mars, and whatever additional costly dreams may arise. Who is to say at this juncture which plan will command a high priority among the many projects currently being debated? However, air traffic is here to stay. Therefore commercial pressures alone will force major improvements in subsonic transport aircraft. This quest for improvements will encourage additional research efforts by industry, NASA, and university laboratories.

These thoughts bring us to the relatively safe ground of taking a look at largely foreseeable aerodynamic and other improvements in commercial airliners. This will be a pleasant return to our subject, since Chapters 3 to 9 have prepared us to follow the technical side of the proposed refinements of current practice.

Advances in the design of airliners deal with different aspects of propulsion and aerodynamics. Again we shall ignore structural improvements, although simplifications of design and the use of lighter materials certainly contribute much to the building of better airplanes. The following discussion is based on the assumption that the speed range of airliners shown in Figure 10.1 and Table 11.1 will remain unchanged into the next century. This is a safe prediction, since all long-haul aircraft currently on order or under design are within this range, as noted before. The primary goal in the coming years will be the reduction of the cost of a seat-mile, the cost to fly one passenger or a unit of weight for one mile. The seat-mile cost of a DC-3 in 1936 was three-and-a-half times higher than that of a DC-10 or B-747 in 1970. This demonstrates that the economics of airliners have been much improved in recent years. In contrast, the seat-mile cost for the Concorde (one of the several reasons that it has not been widely purchased) is roughly that of a DC-3—about three times that of a DC-10—and the number of passengers is much smaller than the number carried by a DC-10.

To lower cost, more efficient propulsion is required: think of the gas mileage of a car. Engine development to increase efficiency is admittedly difficult. Jet turbines must be able to operate at higher temperatures that lead to higher thermodynamic efficiencies. To solve that problem, materials that can stand higher temperatures are required for the moving parts and the enclosures. Closer to aerodynamics, further noise reduction would be a great advantage to make possible the location of airports closer to population centers. Another promising development was mentioned before in Chapter 9.2: new types of propellers for turboprop engines may soon allow an approach to the speed of pure jet airplanes. The turboprop saves fuel; it is already much used to carry freight and in military applications to ferry troops and heavy equipment, including tanks.

Jet engines of the future ought to have a greater tolerance for different grades of fuel. Not all jet fuels at all locations on far-flung flight routes—again the Pacific comes to mind—are refined to the same specifications. It would be practical for scheduling as well as cost-saving for an airliner to be able to fill up anywhere.

Improvements in aerodynamics are essential. For example, drag reduction, which is still possible, would be an important contributor to economic progress. Figure 11.5 gives a breakdown of the various components that make up the total drag of a modern airliner flying at cruising speed. This graph applies to design conditions of a certain wing and is not necessarily valid for all wings. Much of actual flight, however, takes place at different speeds and attitudes and in varying environments, and it is important to improve performance all around. As Hilbig notes, "the classical field of aerodynamics still offers a large performance potential" (see credit for Figure 11.5).

Fuel consumption—everything else being the same—is approximately proportional to the weight divided by L/D, the lift-over-drag ratio. This is a qualitative rather than quantitative statement, of a kind we have encountered previously. Lowering the drag at fixed lift decreases the power requirement and in turn the fuel consumption. Lift and drag are, of course, closely connected, but we are trying to simplify some advanced aerodynamic thinking. The drag due to lift—remember the wing-tip vortices and related effects discussed in Chapter 8.3—can be reduced by various changes in the shape of the wing near its tip.

We also recall aerodynamic adjustment of the circulation, including the Kutta condition, when speed and attitude of an airplane change. The complicated flow field governing the lift distribution, the

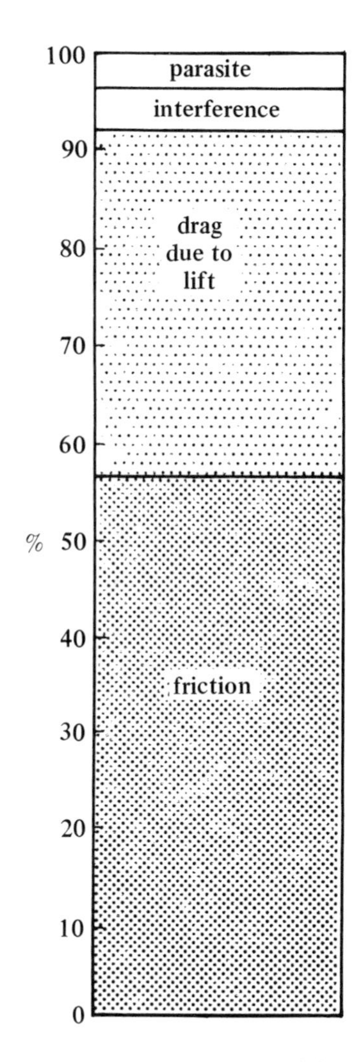

Figure 11.5. *Components of the total drag of a modern airliner. Note the large contribution of surface friction to the total drag.*

downwash, and the interference between wing and fuselage is varied during such a maneuver. For example, supercritical wings are designed for a given cruising speed; in off-design flight their drag increases. Takeoff and landing present still-different aspects. Pilots of current airliners adjust to these conditions by altering the effective wing shape, using the leading-edge devices and the flaps at the back of the wing seen in Figure 8.6. In turn, the aerodynamic coefficients of lift and drag are altered, as indicated highly schematically in Figure 8.4.

Ideally, one desires a continuous, smooth change of wing sections for an airplane, a change that is adapted to any flight condition. Camber (see Chapter 1.4 and Figure A2.2) ought to be adjustable to an optimum shape for takeoff, ascent, cruise, turns, and so forth. Such a possible wing of the future has been called the *intelligent wing*. Its adaptable camber is realized by a combination of moving control surfaces that can be adjusted to provide different shapes, a method that has been tried on fighter aircraft. In particular, such tricks would be employed at the leading edge, whose importance was demonstrated in Figure 8.14. There we noted the drastic variation in pressure very close to the leading edge of the wing. Control of the airfoil shape at that location strongly influences the pressure and in turn the aerodynamic properties of the wing.

The shape of the flexible parts of the wing can be controlled by a computerized system of actuators. The computer would receive the readings of the cockpit instrumentation and of additional devices that measure the acceleration of the airplane, pressure on the wing surfaces, and the like. A computer program could be assembled from wind-tunnel and flight tests to achieve the control found to be most effective for a given flight state.

It may be possible by control of the camber to attain the efficient flow of the supercritical airfoil of Figure 10.5 at a wider range of speeds. This would avoid the strong shock waves and their interaction with the boundary layer shown in Figure 10.4. All these and many additional new schemes are being actively pursued. Their realization, however, is dependent on true similarity of Mach number *and* Reynolds number in model testing. It is expected that the new cryogenic transonic wind tunnels like the one described in Chapter 10.1 will provide the appropriate environment in the near future.

Returning to Figure 11.5, we note that drag due to surface friction caused by the turbulent boundary layer on the airplane accounts for over half of the total drag. The graph applies to a narrow-body air transport, but the situation applies to larger craft as well. Is there any way to reduce the surface friction? Wind-tunnel experiments have demonstrated that very fine grooves engraved in the flow direction on plates do lower by a few percent the turbulent friction with respect to the results shown in Figure 7.7. The details of the drag-reducing surface were copied from the skin of sharks, whose drag in the water has been reduced by evolution to provide rapid swimming. It would be difficult to maintain an aircraft covered with such finely figured surface details, however, because of accumulations of insects, dirt, rain, and snow. But this approach may still be worth pursuing.

Figure 7.7 and the corresponding equations in the text that give laminar and turbulent friction coefficients suggest that laminar boundary layers exert much lower friction at high Reynolds numbers. In fact, at a Reynolds number of 50 million, which is characteristic of a large jet transport at cruising speed, turbulent friction on a smooth flat plate is eleven times that for laminar flow. At this time no solution is in sight to achieve a laminar flow on the fuselage at such Reynolds numbers, which are far above the transition values given in Figure 7.7. Although wings contribute less than half of the total friction drag of Figure 11.5, it may be possible to keep at least part of the wing surface in the laminar regime. Wings with an upper surface that does not depart much from that of the supercritical airfoil shape of Figure 10.5 have been tested. It appears that boundary-layer transition can be delayed in the region of strongly falling surface pressure. Again maintenance problems appear that can trigger the transition, as we saw before. However, a drag reduction for the entire airplane of 15 to 20% seems possible.

In sum, ongoing aerodynamic developments will permit a substantial reduction of the drag, an adaptation of optimal aerodynamic performance to many different flight conditions, a possible reduction of aircraft noise, a lowering of fuel consumption, and, as a consequence of all these things, substantial economic savings. These expected improvements will also add to the safety—for example, by lowering landing speeds—and comfort of the passengers.

At this point we will leave the relatively safe confines of the scientific and engineering approach to commercial flying to look for a moment at problems with air traffic that can be solved only by social and political decision making. So far we have usually seen the airplane in flight, disconnected from the city, the airport, and other air traffic. But nonaeronautical concerns intrude immediately in the real world. The subject of travel time rather than speed of the aircraft is a case in point. It is said that the Concorde cuts travel time in half between New York and Paris. But is that true? Of course not. Traveling to the airport, checking in, waiting to board, sitting on the runway, picking up luggage, going through customs, waiting for ground transportation, and getting stuck in a traffic jam add many hours to the actual flying time. Thus the quicker transatlantic passage of the Concorde can be reckoned only as a fractional saving of the total time of the journey.

The infrastructure of air traffic is an area of activity to which ingenuity equivalent to that of aircraft design must be directed. Airports must be located and designed to greatly simplify all the flight-associated activities that we listed. They should be sufficiently far from population centers so that long runways can be built and noise is no serious problem. At the same time, they must be linked to population centers by networks of high-speed trains on the surface or under ground. Airport traffic must be separated from general traffic, a method that has long been applied in Europe. For example, London's Heathrow has three underground rail stations that get you close to the terminal of your chosen airline. The problem inherent in the long walks or even bus trips in heavy traffic that get you from one airline to another at many airports needs to be solved. It is here that regional interests take over and political decisions are required in cooperation with the federal government and the airlines.

The government is, of course, already involved in many aspects of air traffic, even in the current age of deregulation. For example, it watches over the safety of airplanes, grants permits to newly developed aircraft, operates air traffic control, promulgates airport regulations, and investigates accidents. The deregulation of airlines was carried out to permit greater competition among different carriers. It will be a difficult problem in the future to strike the proper balance between permitting this freedom and ensuring safety while giving the traveler the best possible choices at the lowest cost.

There is a range of currently existing and soon-to-be-available technologies that must be employed by the government and prescribed for the airlines. Air traffic control must be modernized and all-weather landing introduced; collision-avoidance systems and devices for on-board detection of clear-air turbulence, as well as sensors that locate imminent wind shear (strong vertical air currents), must be installed in airplanes as we discussed previously. Some of these technologies are already available; others will be perfected soon.

We noted before that the total number of long-haul airplanes may not increase much in the United States to handle the extended air traffic predicted in Figure 11.2. As a consequence, larger aircraft will handle the traffic, so that the number of takeoffs and landings at a given airport may stay constant or even decrease. Long-range planning far into the next century to handle the necessary changes is essential. There will be external challenges as well. New worldwide routes will be opened up that impinge on the extension of traffic of U.S. airlines. Arrangements will be needed to handle an increased number of foreign airlines in the United States, many more of which will want to fly on what so far have been domestic routes. In 1992 the European Community began to represent a huge unified market. Its airlines may act in concert to extend their service to the Americas and the Far Eastern countries. Increasingly tough competition will arise, and a new stage of commercial flying will appear.

The near future into the next century will be exciting, and one hopes a revival of the old pioneer spirit that conquered the skies will take place to help us master the difficult problems that will arise. Improved airliners must handle increasing traffic. Total travel time—at fixed cruising speed of all airliners—must be shortened. Smaller towns must not be ignored. New urban railroads must connect airports to the population centers. All this has to happen quite aside from aerodynamics, the mature science of flight that will produce many more improvements in airplanes.

Appendix One

Facts from Algebra

In physics and engineering, the laws of nature must ultimately be described by mathematical equations. Such equations are based on observations, and they provide a universal language from which additional results and predictions may be derived. These can in turn be checked by experiment. Our topic—what makes airplanes fly—is based on *mechanics,* the branch of physics which covers *fluid mechanics* and consequently *aerodynamics.* An engineer uses differential equations—for example, the equations of motion (Chapter 5)—and other advanced mathematical methods to describe mechanical processes. A full treatment of these phenomena must be left to the professional engineer. Yet even our objective requires simple mathematics in order to get at least some flavor of the methods of engineering. A purely qualitative or descriptive approach without any numerical values is largely meaningless. What is the drag of a golf ball, the lift of a transport plane and its lift-over-drag ratio, or the surface friction of a submarine without numbers? Engineering requires quantitative answers to questions such as these for even the simplest problems.

The equations we will use are based on elementary algebra, and no more is needed to handle the sample problems in Appendix 5. In this appendix, you will find a brief summary of algebra, including a discussion of logarithms. The latter are essential to the solution of many problems that we encounter in our attempts to understand the world around us. Logarithms are needed to compare atomic scales with those of the universe, to understand the economics of world trade and its implications for you and me, to view bacterial growth or the chemical reactions in a candle's flame, to study the structure of the atmosphere, or to understand the present and future of public transportation. More material on algebra and logarithms than is provided here can be found in high-school textbooks or in encyclopedias on the shelves of any public library.

Symbols

$a + b$, a plus b
$a - b$, a minus b
$a \pm b$, a plus or minus b
$a \times b$, or $a \cdot b$, or $(a)(b)$, a times b

$a{:}b$, or $\frac{a}{b}$, or a/b, a divided by b

$a = b$, a equals b
$a \neq b$, a does not equal b
$a \approx b$, a approximately equals b
$a > b$, a is greater than b
$a < b$, a is less than b
$a{:}b = c{:}d$, a is to b as c is to d
%, percent
∞, infinity

Rules of Arithmetic

$a + b = b + a;\ ab = ba$
$a + (b+c) = (a+b) + c;\ a(bc) = (ab)c$
$a + (b+c-d) = a+b+c-d$
$a - (b+c-d) = a-b-c+d$
$(+a)\ (+b) = +\ ab;\ (-a)\ (-b) = +\ ab$
$(+a)\ (-b) = (-a)\ (+b) = -ab$

$$\frac{+a}{+b} = +\frac{a}{b};\ \frac{-a}{-b} = +\frac{a}{b};\ \frac{-a}{+b} = -\frac{a}{b}$$

$$\frac{a}{b} = \frac{ac}{bc};\ \frac{a}{b} = \frac{a{:}n}{b{:}n};\ \frac{a}{c} \pm \frac{b}{c} = \frac{a \pm b}{c}$$

$$\frac{a}{b} \pm \frac{c}{d} = \frac{ad \pm bc}{bd};\ \frac{a}{c} - \frac{b+d}{c} = \frac{1}{c}(a-b-d) = \frac{a-b-d}{c}$$

Exponents

$a^x a^y = a^{x+y};\ (ab)^x = a^x b^x;\ (a^x)^y = a^{x \cdot y}$

$$a^{-x} = \frac{1}{a^x};\ \frac{a^x}{a^y} = a^{x-y}$$

$$a^{x/y} = \sqrt[y]{a^x};\ a^{1/y} = \sqrt[y]{a}\ \ (\text{e.g., } 4^{1/2} = \sqrt[2]{4} = 2$$

Logarithms

If B is an arbitrarily chosen number greater than unity, then the *logarithm* L of any other number N is defined by

$$N = B^L \text{ or } L = \log_B N.$$

Here $0 < B < \infty$, but $B \neq 1$.

L is called the logarithm to the base B of N. The logarithm of 1 to any base is 0. Values of L for the ranges of N to various bases are listed in logarithmic tables. Such tables can also be used in reverse to find the *antilogarithm*—that is, N can be found for a given L.

The use of logarithms in calculations makes it possible to take advantage of the rules given above for the handling of exponents. Multiplication becomes addition and division becomes subtraction, a situation which is the basis of the slide rule, a device that is rarely in evidence today.

$$\log (a \cdot b) = \log a + \log b; \ \log \frac{a}{b} = \log a - \log b;$$

$$\log a^n = n \cdot \log a; \ \log \sqrt[n]{a} = \frac{1}{n} \log a.$$

With $B = 10$, it is particularly useful to draw logarithmic scales in graphs. (The base 10 is often used, and thus $\log_{10}$ N is usually abbreviated log N.) Wide ranges of order of magnitude can be covered. Recall the internuclear distance in the H_2O molecule of 10^{-10} m. The astronomical unit is the average distance of the earth from the sun, of the order of 10^{11} m. (Its actual value is close to 150 million kilometers.) The logarithms of these two values are –10 and +11 respectively. A graph of just over twenty subdivisions can readily encompass these two extremely different lengths.

Another base of the logarithm, B = e = 2.718 . . . , leads to *natural* logarithms, abbreviated by $\log_e$ N = *ln*N. In the present context, we need not be concerned with the definition of the so-called transcendental number "e"—an unending fraction—or the mathematics which makes natural logarithms an important tool (see Chapter 4).

It is unlikely that one would use an extended logarithmic table these days. Such tables are already included in the circuitry of simple hand-held calculators. However, it is useful to compare the values in a table with results of calculator computations. In the latter case the mathematical content is internal and hidden from view, and possibly from understanding. It is not to be inferred from these remarks that I am suggesting a return to learning logarithmic tables by heart!

Logarithms were not known to ancient Arabic and Greek mathematicians, but rather they are creations of the sixteenth century. The name of John Napier (1550–1617), who invented logarithms, is attached to the natural or Napierian logarithm, the logarithm based on the number e. Henry Briggs (1561–1630), a professor of geometry at Gresham College and later at Oxford University, subsequently introduced the system of logarithms in base ten, and his name became attached to it. Briggs calculated a logarithmic table for 30,000 numbers to fourteen places, a remarkable feat in his time.

Finally, I list some useful formulas for geometric calculations. These equations show general relationships that are handy for calculations of area and volume.

Useful Formulas for Geometric Figures

Symbols:

d, diameter; *r*, radius; *A*, area; *c*, circumference.

V, *volume; h*, height; $\pi = 3.14159 \ldots$. (using 3.14 is sufficient for engineering calculations).

Circle:

$d = 2r;\ c = 2\pi r = \pi d;$

$$A = \pi r^2 = \frac{\pi d^2}{4} = \frac{c^2}{4\pi}.$$

Sphere:

$A = 4\pi r^2 = \pi d^2.$

$$V = \frac{4\pi r^3}{3} = \frac{\pi d^3}{6}.$$

Right circular cylinder:

$A = 2\pi rh$ for the curved surface, $V = \pi r^2 h$.

Appendix Two

Model Testing and Similarity

A2.1 Aircraft Nomenclature and Model Testing

Before introducing the main subject of this appendix, we should define a few terms relating to aircraft components. In Figure A2.1, we see a schematically drawn twin-engine monoplane, a type of aircraft characteristic of current airliners built in many countries. The undercarriage—the apparatus of wheels on which the airplane starts and lands—is retracted; the craft is in the cruising mode. Various external parts are labeled, with particular emphasis on the control devices. Note that the wing warping by cable invented by the Wright brothers has been replaced by ailerons, small control surfaces located near the wing tips.

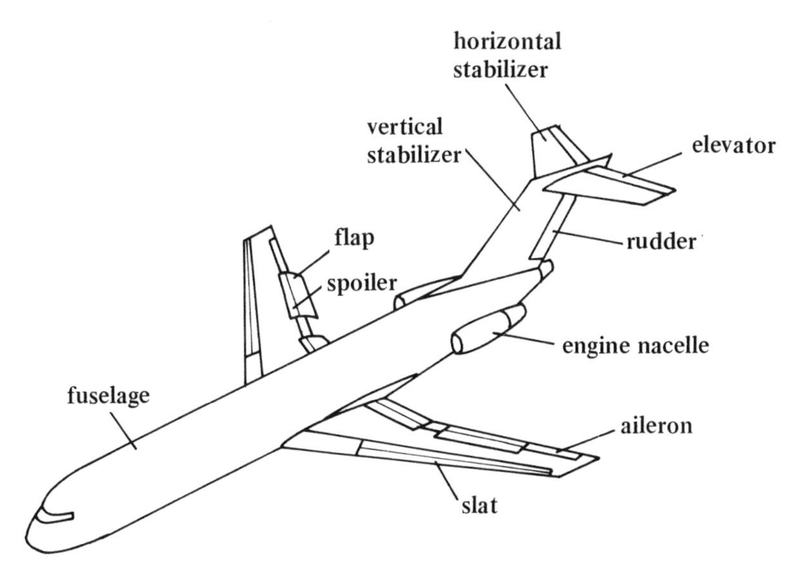

Figure A2.1. The parts of an airplane.

Table A2.1. The Greek alphabet

Α	α	Alpha
Β	β	Beta
Γ	γ	Gamma
Δ	δ	Delta
Ε	ε	Epsilon
Ζ	ζ	Zeta
Η	η	Eta
Θ	θ	Theta
Ι	ι	Iota
Κ	κ	Kappa
Λ	λ	Lambda
Μ	μ	Mu
Ν	ν	Nu
Ξ	ξ	Xi
Ο	ο	Omicron
Π	π	Pi
Ρ	ρ	Rho
Σ	σ	Sigma
Τ	τ	Tau
Υ	υ	Upsilon
Φ	φ	Phi
Χ	χ	Chi
Ψ	ψ	Psi
Ω	ω	Omega

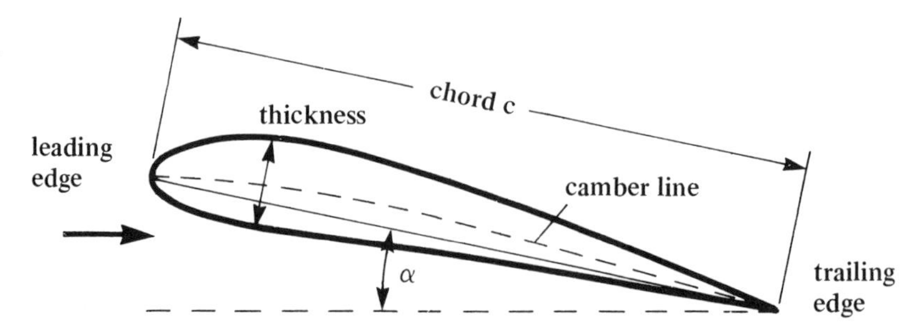

Figure A2.2. Terms used to describe a wing section.

Figure A2.2 gives the special terms that are applied to wings. A slice called a *wing section* is shown, with labels applied to its parts. Of particular importance in the design of a wing or *airfoil* is the *camber.* The dashed camber line defines the shape; it lies exactly midway between the upper and lower wing surfaces. By governing the details of the thickness distribution that determines the external shape of a particular design, camber determines the wing's aerodynamic characteristics such as lift and drag. The chord is defined as the distance between the leading and trailing edges. It is indicated by the solid chord line. Locations on the wing, such as those of pressure taps used in wind-tunnel experiments, are often indicated in graphs by a dimensionless value (see the next section). This value is arrived at by dividing the actual distance from the leading edge, x, by the chord, c, to get x/c. Figure 8.14 is a graph of this type. The angle of attack, α (alpha), is the angle between the chord line and the flight direction (see Table A2.1 for the Greek alphabet, which is occasionally used for symbols). In a wind tunnel the angle of attack indicates the attitude with respect to the oncoming wind. In calm air, the large arrow in Figure A2.2 indicates the horizontal plane.

In Figure A2.3 we see some of the many ways in which an airplane designer can put wings and fuselage together. This great variety is found today mostly in the smaller aircraft of *general aviation,* a term that includes private airplanes. Most big jet airliners are low-wing monoplanes with swept-back wings. This wing shape is required at the high speed of modern airliners for reasons that are discussed in Chapter 10.1. However, there remains a remarkable variation in the arrangement of the two, three, or four jet engines that push an airliner through the air.

Model testing is at the heart of aerodynamics. Before an aircraft, a submarine, the superstructure of a ship, or a skyscraper is actually built, scale models are exposed to airflows to examine the aerodynamics of the new design. For airplanes in particular, the lives of test pilots are at stake, not to mention large expenditures of time and money. The fundamental problem is to determine the exact conditions of the two flows about model and full-scale prototype that insure *similarity.* Similarity implies identity of the structure of the two flows so that one can apply the results of the model test to the full-scale object. It is this topic that we addressed in Chapter 6; here we shall derive the physical arguments that led to the introduction of the Reynolds number, the primary dimensionless similarity parameter of aerodynamics, and list other dimensionless parameters.

The first testing facility was the whirling arm, a machine not unlike a merry-go-round on which a model was spun on a circular track. This device was invented in 1746 by Benjamin Robins, a British artillery engineer. Robins—in an additional original experiment—was also the first to measure the drag of spheres at supersonic speed by firing cannonballs into a ballistic pendulum, with which he measured their impact. These beautiful experiments with supersonic flow preceded similar work by over fifty years. Sir George Cayley, whom we previously encountered as the originator of the modern airplane configuration, performed the first whirling-arm tests on airfoils around 1800.

Wind tunnels—ducts in which a controlled, uniform airflow is blown against a scale model—are over one hundred years old. Names that we have also encountered before, such as Mach, Joukowski, the Wright brothers, Prandtl, and Eiffel, were connected with this development. In addition, many other testing schemes appeared; among them we find flight tests of models attached to aircraft, free flight of models dropped from aircraft, tethered flight (remember the Wright gliders), firing ranges, models mounted on high-speed rocket sleds on rails, and water tunnels. All these methods have their place in the spectrum of aerodynamic research. However, the wind tunnel remains the most common tool in all speed ranges.

Let us now concentrate on conditions under which an experiment on a small model yields results that are applicable to a full-scale airplane. To prepare the ground for such a discussion, we must first work our way through a number of important concepts. These concepts have at first hearing an abstract ring to them, yet their application is eminently practical, as we shall see.

A2.2 Dimensions and Units

In physics and engineering—including aerodynamics—observation and measurement are the basis of our insights. We are concerned with relationships of physical quantities, and these physical quantities have *dimensions:* here, we shall need primarily the following dimensions:

Length L
Mass M
Time T.

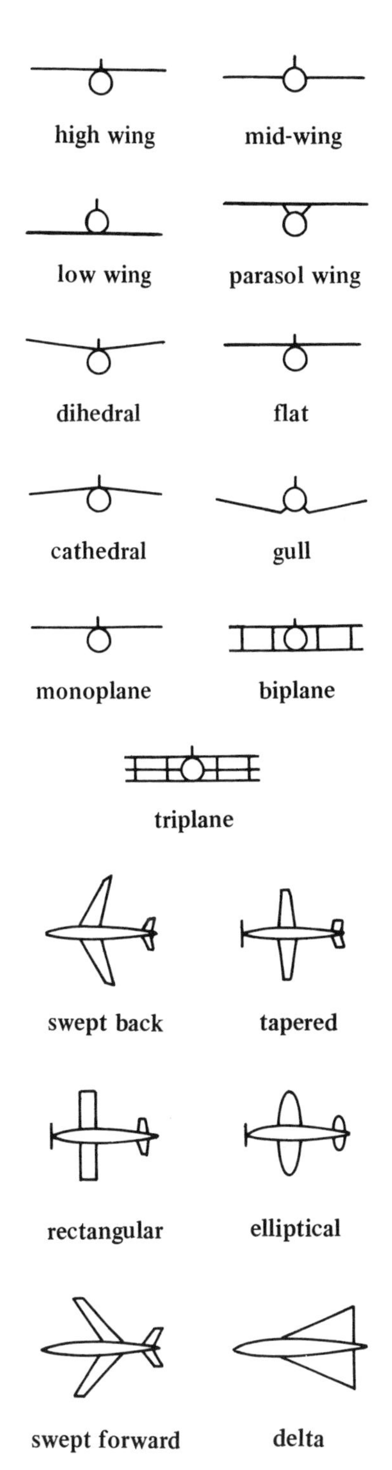

Figure A2.3. *Aircraft configurations.*

These fundamental dimensions permit the expression of various other dimensions or physical quantities, such as area, volume, speed, power, and density. Occasionally—as in the discussion of the atmosphere or high-speed aerodynamics—another fundamental dimension, temperature, appears.

Whenever we write an equation that expresses a law of physics, each term in the equation must have the same dimension or combination of dimensions. For example, the mass of an airplane may be computed by

$$\text{total mass} = A + B + C + \ldots,$$

where A, B, C, and so forth stand for the mass of the fuselage, the engine, and other parts of the plane. Each term has the dimension of mass, M, and the equation is *dimensionally homogeneous*. Dimensional homogeneity is a general requirement. You can't mix apples and oranges in a single equation.

Similarly, we may wish to add the forces acting on an automobile to retard its motion. Such forces are the internal friction of mechanical parts, the rolling friction arising from the contact of the tires with the road surface, and the aerodynamic drag or resistance of the air to the car. The dimensions of these forces must of course be identical. Newton's second law states that force equals mass times acceleration, with the acceleration as the rate at which speed changes over time. Since the dimension of speed is L/T, acceleration is characterized by L/T^2. In turn, the dimension of force is ML/T^2 or MLT^{-2} (see Appendix 1); therefore the dimension of force, F, contains only length, mass, and time. In the solution of engineering problems it is often useful to use force rather than mass.

Returning to the summation of the forces resisting the forward motion of our car, we must now make sure that all terms in the equation have the *dimension* of force, MLT^{-2}, to assure dimensional homogeneity. Also note that these steps must be taken *prior* to a decision as to which system of *units* (British or metric and corresponding subsystems) to use. The dimensional approach clarifies the physics of the problem in the selection of the relevant parameters.

Most physical quantities that are needed in aerodynamics can be derived from the three basic dimensions of mechanics given above. Since aerodynamics is based on fluid mechanics (statics and dynamics), aerodynamic parameters are in turn taken from that field. Quantities and dimensions of parameters important to aerodynamics are listed in Table A2.2, where—as previously in the text—the Greek alphabet is used (see Table A2.1). This provides a set of symbols to ease the confusion of nomenclature.

In what follows we will assume that the reader is familiar with the basic ideas and laws of mechanics, such as Newton's laws of inertia, which require an understanding of the definition of mass. The

Table A2.2. Symbols and fundamental dimensions of the physical quantities required for an understanding of aerodynamics.

Symbol	Physical Quantity	Dimension
A or S	area or surface	L^2
V	volume	L^3
U	velocity (or speed)	LT^{-1}
a	acceleration	LT^{-2}
Q	volumetric flow rate	L^3T^{-1}
m	mass flow rate	MT^{-1}
F	force	MLT^{-2}
W	work (also E, energy	ML^2T^{-2}
P	power	ML^2T^{-3}
p	pressure (force/area)	$ML^{-1}T^{-2}$
ρ (rho)	density (mass/volume)	ML^{-3}
μ (mu)	dynamic viscosity	$ML^{-1}T^{-1}$
ν (nu)	kinematic viscosity	L^2T^{-1}

second law—force equals mass times acceleration—and the definition of work—force times distance—must be understood. In addition, energy with the dimension of work and power defined as the work done per unit time are needed. These shorthand definitions are insufficient for a deeper understanding, and readers not familiar with this part of physics are referred to the many textbooks in that field or the relevant chapters in encyclopedias of science such as those listed in Appendix 4.

Once we have verified the dimensions of a given expression, such as Bernoulli's equation (Chapter 5), we can append *units* in order to provide numerical results. In aerodynamics—as in all branches of engineering—we must be able to give a numerical value to the lift of a wing or the drag of a ship. Return again to the total force retarding the motion of an automobile. This can be expressed in the British force unit the pound (lb, or occasionally lb_f) or the metric force unit the newton (N). The choice of units is immaterial, but the identical force unit must be used for each component of the total drag. If the components were originally expressed in different units, they must be converted to a common unit. Conversion factors to accomplish this for the British and the SI systems are listed in Appendix 3. In spite of the differences between the two systems, the unit of time, the second (s), is fortunately identical!

A2.3 Similarity Parameters and Model Testing

Let us return to the question originally posed: How do we perform wind-tunnel experiments on scale models whose results are applicable to the full-scale prototype? In Chapter 6.1 we saw that the key to this problem is testing the model at the identical Reynolds number of the prototype, the real airplane. Here we will go beyond the discussion of Chapter 6.1 and derive the Reynolds number by *dimensional reasoning.*

This derivation is more difficult to follow than the remainder of the text. We address readers who wish to understand the characteristic method by which engineers tackle such problems. *Dimensional analysis,* the basis of modeling physical phenomena, is among the original contributions to science made by engineers. The common idea that scientists find out about nature and engineers apply their results to the design of the technological world is a fallacy. Engineers often contribute to basic physics and chemistry, and so-called basic scientists often work as excellent engineers in the design of experimental apparatus. Dimensional analysis is a major contribution to knowledge by engineers.

How do we go about dimensional analysis? For a given problem we attempt to identify the different physical parameters that govern the particular flow involved. You expect, for example, that the size of an automobile will have an effect on its aerodynamic drag, while the power of the engine is immaterial in this context. Let us take an airplane that flies at a low speed far removed from the speed of sound. As with the car, the size of the aircraft will obviously be important to determine forces such as lift and drag. Surely the speed of the aircraft will have an added effect on the aerodynamic forces. Note that size and speed are chosen by us for a *given* aircraft. The airplane flies at some altitude in the atmosphere, whose *properties* at that level affect the flow field. Air density—the air mass per unit volume—and air viscosity enter. The viscosity, the internal friction that we encountered in the discussion of the boundary layer, is a function of air temperature, and the sheer forces between neighboring fluid layers depend on it.

We have now determined intuitively that the flow about a body moving at low speed fully immersed in a fluid* is governed by the size of the body, its speed, and the fluid's properties of density and viscosity. Indeed, physical insight is the basis of our reasoning; dimensional analysis is as much an art as a science. In our discussion we thought of an airplane, but the same rules apply to a submarine or a fish, a golf ball or an arrow.

These considerations are by definition equally valid for a scale model in a wind tunnel and the actual aircraft flying through the air. We now ask ourselves what conditions must be met to assure that the flows about these two objects are identical. Obviously model and prototype must be geometrically similar—that is, they must look exactly alike. Streamline patterns about both—marked by dye or smoke—must also look alike, with the only difference being a difference in scale. The flows so established are called *similar.* We now search for a *similarity parameter*—a combination of variables expressing length, speed, and fluid properties—that assures the identity of the two flows. To render this condition certain, it is stipulated that the *ratio* of the forces acting on some fluid element in the flow field about model and aircraft must be identical for both.

The variables governing the forces have been identified, and it remains to find a suitable combination of them that is *dimensionless.* For one, inertial forces represented by the energy of the fluid moving around the object (recall Bernoulli's theorem) act on the fluid particle. In turn, viscous forces counteract those of inertia. Depending on the relative values of these two forces, fluid motion may be dominated by viscosity—like honey running off a spoon—or inertial forces

---

*A ship, on the other hand, operates on a water surface, causing waves to develop. Here our thoughts do not apply, while they are valid for a submarine below the surface.

may win out to topple the orderly layers of fluid and initiate turbulent motion (see Chapter 5).

Take our four variables governing incompressible flow. The inertial forces are given by Newton's law of force (force = mass × acceleration). With the characteristic length, l, dominating the scale, and recalling the definition of density, ρ, we find ρl^3 for the mass of a fluid element taken as a cube. The acceleration is given by the change of flow speed U over the length of the model. This acceleration—from calculus we have dU/dt—takes place during the time interval l/U. Consequently, we find for the acceleration $U/(l/U) = U^2/l$. (It is useful for you to write out the dimensions of this and the following expressions to check the calculation.) Finally, by this reasoning, we have for the inertial forces acting on an element of fluid

$$f_i \sim \frac{\rho l^3 U^2}{l} = \rho l^2 U^2.$$

(The symbol "~," which represents a proportionality, may be used in lieu of "approximately.")

Next we express the viscous force exerted by the shear due to internal friction—that is, the shearing force per unit area. For our fluid element, this area is given simply by its side, l^2. The shear force is linked to the viscosity, as we saw in the discussion of boundary layers. We thus find for the viscous force

$$f_v \sim \tau l^2 = \mu \frac{U}{l} l^2 = \mu U l,$$

making use of Newton's law of shear (Chapter 5).

Finally we determine the ratio of inertial to viscous forces acting on the fluid element by setting f_i/f_v, to obtain from the two previous expressions

$$\frac{\rho l^2 U^2}{\mu U l} = \frac{\rho l U}{\mu}\left[\frac{\mathrm{M}}{\mathrm{L}^3}\mathrm{L}\frac{\mathrm{L}}{\mathrm{T}}\frac{\mathrm{LT}}{\mathrm{M}}\right].$$

In the square brackets we have inserted the proper dimensions for the variables making up the force ratio, and we note that this final expression is indeed *dimensionless*—that is, all the dimensions cancel. You can verify this fact by using Table A2.2. We have now arrived at an expression governing flow fields about geometrically similar bodies in the absence of gravity. Gravity forces in air flows are unimportant (see Chapter 5). In heavier fluids we can simply counteract gravity by buoyancy, a requirement met, for example, by a submarine or a diver. The derived ratio of inertial to viscous forces was first experimentally explored by Osborne Reynolds (Chapter 5); it is now called the *Reynolds number* in his honor, and is designated by *Re*:

$$Re = \frac{Ul\rho}{u} = \frac{Ul}{v}.$$

In the last term of the equation, we have introduced the kinematic viscosity, v, as the ratio of viscosity to density. Its value is constant in incompressible fluid flow. This results from the fact that at fixed temperature, density and dynamic viscosity—previously called simply viscosity—are constant. The choice of the characteristic length l is arbitrary. However, if the length is taken for the chord of an airfoil for both model and prototype, the Reynolds number is fully determined. To repeat, if the Reynolds numbers of the flow about prototype and scale model are identical, the two flow fields are identical and the results of model experiments can be applied to the full-scale object. This fact is the basis of all aerodynamics testing.

The definition of the Reynolds number permits a wide range of variations of individual parameters to achieve similarity. For example, for a 1:10 scale model, the wind-tunnel flow speed must be ten times the flight speed of the original in order to provide identical Reynolds numbers. In turn, a wind tunnel can be pressurized to increase the density, a fact which can be used to increase the Reynolds number of the model at fixed airspeed. Other fluids with different kinematic viscosities can even be used. Testing of small models in water, whose density is about eight hundred times that of air and whose kinematic viscosity is about seven one-hundredths that of air, can duplicate large-scale airflows and their high Reynolds numbers. The newest designs for wind tunnels approaching the speed of sound—transonic wind tunnels—depart from all previous schemes, operating with pure nitrogen rather than air. The nitrogen is kept near its boiling point of about –200°C (–330°F), since at such low temperatures the viscosity of a gas becomes very low. Everything else being the same, the model's Reynolds number will therefore increase.

These ideas discussed here have provided one of the bases of the enormously rapid development of aviation and other applications of aerodynamics. Clearly, great economic advantages accrue from designing, building, and altering small models in the course of the development of a new aircraft. This is much more economical than rebuilding the full-scale article, and safety can be maintained during the development process. Although the Wright brothers had no formal technical training, they were sound engineers who knew about similarity, and they tested their model airfoils in a small wind tunnel of their own construction. From this accomplishment to the "flying" motion of a modern submarine, the development of a heart-lung machine, and the safe return of the space shuttle to the earth, the Reynolds similarity law has dominated fluid mechanics and its subfield aerodynamics. The range of Reynolds numbers encountered in aerodynamics is enormous, spanning something like ten orders of magnitude, from settling dust particles to large airplanes and dirigibles, as we saw in Chapter 6.

Finally, we must note that the Reynolds number is not sufficient as a single similarity parameter in all situations. We must add additional considerations if the flow speed approaches the *speed of sound.* Motion at such a clip forces us to consider the air to be compressible—that is, the pressure changes in the flow are sufficiently substantial to alter the density of the air. In the flight regime approaching and exceeding the speed of sound (340 m/s = 1,224 km/h = 761 mph at sea level), the laws of fluid mechanics become more involved, departing from those of liquids and low-speed air. A new dimensionless parameter becomes important: the *Mach number,* named after the Austrian physicist and philosopher Ernst Mach (1838–1916). This parameter is defined by

$$\mathrm{M} = U/a,$$

the ratio of the flow speed to the speed of sound. Dividing one speed by the other obviously produces a dimensionless parameter. The value of the Mach number determines two entirely different regimes of flow. For $\mathrm{M} < 1$ we speak of subsonic flow; if $\mathrm{M} > 1$, we have supersonic flow, where the object moves at a speed faster than that of sound. The body outraces its own sound field—or noise—a condition leading to remarkably different flows, as discussed in Chapter 10. But the effects of compressibility—and consequently the need to consider M in addition to *Re* in model tests—are not important at Mach numbers that are roughly below $\mathrm{M} = 0.3$, or speeds of about 230 mph.

In wind tunnels we measure forces or pressures on models, and the question arises how we can apply these experimental results to full-scale airplanes. For one, as we have just seen, we work at identical Reynolds numbers to produce dynamic similarity of the flows about model and prototype. Next we recall from Chapter 1 that the resultant air force—a vector—acting on the center of pressure may be resolved into two components. One is counted in the direction of the motion—the direction of the wind blowing against the model in the wind tunnel—the other is counted at right angles to the first. The resulting two vectors represent the *drag force,* D, and the *lift force,* L. The drag impedes forward motion, giving the total air resistance. The lift force opposes the force of gravity; it must be equal to the weight in order to sustain level flight. The power of the aircraft computed for cruising conditions is based on these two forces acting at a steady flight speed. Drag and lift can be directly measured in newtons (or pounds) by a wind-tunnel balance from which the model is suspended. The balance system (see Figure 6.1) is designed to permit a variation of the angle of attack, α, of the airfoil, the angle between the chord of the wing and the direction of flight (see Chapter 1).

Further analysis of the forces and their dependence on the variable of a given flow suggests additional simple dimensionless expressions. We write for the *drag force*

$$D = c_D\, Sq,$$

and for the *lift force*

$$L = c_L\, Sq,$$

where c_D and c_L are newly introduced *dimensionless coefficients*. Rewriting the equations for the coefficients, we have

$$c_D = D/(Sq),$$

and

$$c_L = L/(Sq).$$

In these equations S is an agreed-upon area. For the drag, the cross section of a ball or of an automobile may be used. For the lift, the surface area of a wing is often taken to define S.

As before, the choice of length is somewhat arbitrary, provided that we apply it to the same location in model and prototype. The dimension of S is L^2, and q is the dynamic pressure

$$q = (\rho/2)\, U^2,$$

a term that appeared in Bernoulli's equation (Chapter 5). Using our table of dimensions, we find that q has in fact the dimension of pressure. Therefore we have the dimension of force for the product of area times pressure as required in the denominator of the equations for drag and lift coefficients. Since drag and lift are forces, we note that the coefficients are indeed dimensionless. (To sharpen the perception of units, it is useful to go through the same calculations using the British and SI systems.)

We can now perform typical aerodynamic experiments in the following sequence. We wish to test the drag of a small airplane and build a scale model of the machine. With flight speed and the known air properties at a given altitude, we compute the Reynolds number of the airplane. By a suitable choice of speed in relation to model size, or a pressure change, we duplicate this Reynolds number with the model in the test section of a low-speed wind tunnel. Next we measure the drag force as a function of the angle of attack with the wind-tunnel balance. This force function is made dimensionless by division by the product of area and dynamic pressure. Repeating this experiment at different Reynolds numbers, we establish the function $c_D\,(Re,\alpha)$. The lift force is treated similarly, and the actual drag and lift for the prototype may now be computed from the coefficients and the flight conditions.

Identical methods apply to aerodynamic experimentation with

models of automobiles, trains, superstructures of ships, buildings, and the like. Even for vehicles such as cars, the aerodynamic lift is important, since one wishes to remain earthbound! If we deal with compressible flow, the drag function becomes more complicated. In addition to the Reynolds number, the similarity laws for compressible flow fields require that the Mach number be duplicated. This is a difficult matter indeed, since it is not easy to duplicate both similarity parameters simultaneously in a single wind-tunnel experiment; complete simulation is in fact often impossible. However, once the boundary layers on model and prototype become turbulent, effects of the Reynolds number are less pronounced, and the Mach number becomes the primary similarity parameter.

Finally, we are often interested in the distribution of forces on the roof of a building or at different locations on an airplane wing, so that we can design the structure of the roof or wing properly. To help the structural engineer, the aerodynamicist measures the *distribution of pressure* on a model,* and by analogy to our prior functions, we can reduce the variables entering the experiment to a single *pressure coefficient,* c_p. This coefficient is defined by

$$c_p = \frac{p - p_\infty}{q},$$

again with $q = (\rho/2)\, U^2$. In the equation for the pressure coefficient, p is the locally measured pressure, and p_∞ is the *free-stream pressure* ahead of the model at a location not affected by it. The pressure coefficient may be positive or negative, depending on whether the value of p is above or below that of the undisturbed flow. Moreover, $c_p = 1$ at the stagnation point of the model—its foremost point—where the local flow speed is zero. Also, $c_p = 0$ if $p = p_\infty$, that is, if the local pressure on the model is equal to that of the free stream. The pressure can be positive or negative, depending on the pressure on the model measured in relation to the free-stream pressure. These relations are readily demonstrated for an airfoil, as shown in Figure 8.14.

We have concentrated on the most important dimensionless coefficients that describe aerodynamic performance. However, the aerodynamicist utilizes several additional coefficients applying to the moments, normal forces, and the like (see Chapter 9.1), all based on the same reasoning. They are important to determine the aerodynamics of control surfaces and they provide stability calculations, but they will not concern us here.

---

*For this purpose, small holes are drilled in the model's surface, and pressure gauges are attached via tubing.

Appendix Three

History of the Metric System: The SI System and Conversion Tables

The International System of Units (officially called *Système international d'unités* in French, and designated SI in *all* languages) has been adopted by the principal industrial nations of the world, including the United States. This convention of metrology—the science of weights and measures—has evolved from a definition of a unit of length, the meter (m), and it is the latest form of the metric system, which originated in revolutionary France. The unit of mass, the kilogram (kg), was created by members of the Paris Academy of Sciences, and adopted by the French National Assembly in 1795. John Quincy Adams, at that time the twenty-eight-year-old U.S. minister to the Netherlands appointed by George Washington, hailed this event as "rare and sublime."

In retrospect, it seems unfortunate that, in spite of such admiration, the recently freed British colonies did not—for plausible reasons—take the opportunity to join in this and other agreements on international standards. Congress finally legalized the use of the metric system throughout the United States on July 28, 1866. Since April 5, 1893, all legal units of measure used in the United States have been metric units or exact numeric multiples of metric units. The calibration of the scale in the local grocery store is based in the last analysis on the mass of a small metal cylinder deposited in the basement of a government building in Paris. This little cylinder of the mass of one kilogram represents the only defining object of the original metric system that has survived to this day. (The U.S. National Bureau of Standards is a guardian of an exact copy of it.) The standard for the meter, on the other hand, is now given by a certain number of wavelengths of a given radiation, rather than by the length of a particular meter stick.

Despite the legal adoption of the metric system, it is remarkable how little we encounter it in the daily life of the United States. We are indeed the remaining stronghold of the British system, a system that is crumbling rapidly even in the country of its origin. Science, engineering, and world trade depend on internationally recognized units, and we can no longer remain behind. All of us will have to get used to this system, and therefore it is essential to develop a "feel" for metric quantities.

Once one is over the initial hurdles, the ease with which calculations can be made becomes apparent. Just like our currency, the metric system uses the decimal system. This subdivision of weights and measures by the factor ten, although already known to the ancient Sumerians, was suggested by the Dutch mathematician Simon Stevin (1548–1620) and promoted in the late seventeenth century by Gabriel Mouton, vicar of St. Paul of Lyon. Mouton's reckoning led to great simplification—for example, in the conversion of linear measure to area or volume. Very large and very small numerical values can be expressed and manipulated with ease using powers of ten (see Ap-

TABLE A3.1. NAMES AND SYMBOLS OF SI BASE UNITS AND SI-DERIVED UNITS.

Quantity	Name of Unit	Symbol and Units	
	SI Base Units		
length	meter	m	
mass	kilogram	kg	
time	second	s	
	SI-Derived Units		
area	square meter	m^2	
volume	cubic meter	m^3	
frequency	hertz	Hz	s^{-1}
mass density (density)	kilogram per cubic meter	kg/m^3	
velocity, speed	meter per second	m/s	
acceleration	meter per second squared	m/s^2	
force	newton	N	$kg(m/s^2)$
pressure (mechanical stress)	pascal	Pa	N/m^2
kinematic viscosity	square meter per second	m^2/s	
dynamic viscosity	newton-second per square meter	$N(s/m^2)$	
work, energy, quantity of heat	joule	J	Nm
power	watt	W	J/s

pendix 1): the astronomical unit—the distance of the earth from the sun—is roughly 10^{21} (a one followed by twenty-one zeros!) times longer than the diameter of a water molecule; the factor 10^{21} expresses 21 *orders of magnitude.*

But now let us turn to the units themselves. Although I use the SI system in this text to promote understanding, I often resort to British units, since a better intuitive perception is attached to them. In problem solving, finally, as in the case of dimensions, every term in an equation must have the same unit or combination of units for the result to be physically consistent and numerically accurate. Various names, units, and symbols are given in Tables A3.1 to A3.3.* Table A3.1 contains the SI units and units derived from them that are relevant to our interests. Table A3.2 gives prefixes, multipliers that are expressed in powers of ten (Appendix 1) in order to provide shorthand designations for small or large values. The kilogram, for example, is one thousand grams—that is, 1 kg = 10^3g, the prefix kilo implying 10^3. In turn a milligram—one thousandth of a gram—is given by 1 mg = 10^{-3}g. The prefixes M and μ (mu), which are often encoun-

Table A3.2. SI prefixes to handle large and small numbers.

Factor by which unit is multiplied	Prefix	Symbol
10^{12}	tera	T
10^9	giga	G
10^6	mega	M
10^3	kilo	k
10^2	hecto	h
10	deka	da
10^{-1}	deci	d
10^{-2}	centi	c
10^{-3}	milli	m
10^{-6}	micro	μ
10^{-9}	nano	n
10^{-12}	pico	p
10^{-15}	femto	f
10^{-18}	atto	a

Table A3.3. Conversion factors listed alphabetically by physical quantity.

To Convert from	to	Multiply by
Acceleration		
foot/second^2	meter/second^2	–01 3.05
Area		
foot^2	meter^2	–02 9.29
hectare^2	meter^2	+04 1.00
inch^2	meter^2	–04 6.45
mile^2 (U.S. statute)	meter^2	+06 2.59
Density		
gram/centimeter^3	kilogram/meter^3	+03 1.00
slug/foot^3	kilogram/meter^3	+02 5.15
Energy		
erg	joule	–07 1.00
foot lbf	joule	+00 1.36
kilowatt hour	joule	+06 3.60

(continued on next page)

*All values shown are taken from E.A. Mechtly, *The International System of Units,* NASA SP-7012, Scientific and Technical Information Office, National Aeronautics and Space Administration, Washington, D.C.

Table A3.3. (Continued)

To Convert from	to	Multiply by
Force		
dyne	newton	−05 1.00
lbf (pound force, avoirdupois)	newton	−00 4.45
Length		
angstrom	meter	−10 1.00
foot	meter	−01 3.05
inch	meter	−02 2.54
micron	meter	−06 1.00
mil	meter	−05 2.54
mile (U.S. statute)	meter	+03 1.61
mile (international nautical)	meter	+03 1.85
yard	meter	−01 9.14
Mass		
gram	kilogram	−03 1.00
lbm (pound mass, avoirdupois)	kilogram	−01 4.54
ton (metric)	kilogram	+03 1.00
ton (short, 2,000-pound)	kilogram	+02 9.07
Power		
calorie (thermochemical)/second	watt	+00 4.18
horsepower (550 foot lbf/second)	watt	+02 7.46
horsepower (metric)	watt	+02 7.35
Pressure		
atmosphere	newton/meter^2[a]	+05 1.01
bar	newton/meter^2	−05 1.00
inch of mercury (32°F)	newton/meter^2	+03 3.39
lbf/inch^2 (psi)	newton/meter^2	+03 6.89
millimeter of mercury (0°C)	newton/meter^2	+02 1.33
Speed		
foot/second	meter/second	−01 3.05
kilometer/hour	meter/second	−01 2.78
knot (international)	meter/second	−01 5.14
mile/hour, mph (U.S. statute)	meter/second	−01 4.47
Viscosity		
stoke (kinematic viscosity)	meter^2/second	−04 1.00
foot^2/second (kinematic viscosity)	meter^2/second	−02 9.29
lbf second/foot^2	newton second/meter^2	+01 4.79
poise	newton second/meter^2	−01 1.00
Volume		
barrel (petroleum, 42 gallons)	meter^3	−01 1.59
foot^3	meter^3	−02 2.83

[a]The unit N/m^2 is called the pascal, abbreviated by Pa.

tered, express respectively the multiples of one million or one-millionth of some unit. (The astronomical unit is given by approximately 1.5×10^{11} m [150 million km], but no specific prefix has been assigned to 10^{11}, or conversely to 10^{-11}.)

In Table A3.3 conversion factors are given for quantities in the SI system from the British system. Mechtly states, "The table expresses the definitions of miscellaneous units of measure as exact numerical multiples of coherent SI units, and provides multiplying factors for converting numbers and miscellaneous units to corresponding new numbers and SI units." We have again selected those of the many entries that apply to our field, with a few units of general interest added. If we calculate to the two significant decimal places given, the results will have a higher accuracy than those obtained in the best aerodynamic experiments we can devise.

The multipliers on the right-hand side of the table list the positive or negative power of ten to be multiplied by the factor that follows. For example, –02 means 10^{-2} and +02 means 10^{2} to be multipled by the factor that follows. Take the height of Mount Everest as 29,029 ft. In the section for length in Table A3.3, note that –01 3.05 as printed means $10^{-1} \times 3.05 = 0.305$, the number by which the height of Mount Everest expressed in feet must be multipled to find the height in meters. Therefore we calculate $29{,}029 \times 0.305 = 8{,}854$ m. Conversion of numbers given in meters to feet is done by *division* by the multiplier.

It is important to get used to the SI system by developing a feel for the magnitude of various units; simply remember that one foot is close to three-tenths meter or that three feet are about one meter. The table is particularly useful for combined units, such as values given for density, viscosity, and the like, to compute parameters such as the Reynolds number. In the section for viscosity, both dynamic and kinematic viscosity appear. From Table A2.1 we find for the dynamic viscosity the dimension $ML^{-1}T^{-1}$, and for the kinematic viscosity L^2T^{-1} applies. In Table A3.3, therefore, the SI units Ns/m^2 and m^2/s are to be used respectively for the two viscosities. Property values of fluids are found in handbooks; they are often given in metric units.

To convert degrees Fahrenheit (T_F) to degrees Celsius (T_C; formerly often called centigrade), use the formula $T_C = (5/9)(T_F - 32)$. The factors 5/9 and 9/5 (if you reverse the procedure) are given by 0.56 and 1.80 respectively. A calculator will assist you in performing this and other conversions. However, to get used to the system it is important to remember rough values (for example, 1 mi = 1.6 km) rather than worry about extreme numerical accuracy. In the case of temperatures, memorize some obvious pairs of values that you encounter in daily life. We find 32°F = 0°C; 50°F = 10°C; 75°F = 24°C; 98.6°F= 37°C; and 212°F = 100°C.

The last word concerns the use of the pound and variants of it with different meanings. In the text and the homework problems, we use the pound, abbreviated lb, consistently as pound force. This unit is shown as lb_f in the table to avoid confusion with several other units often found in technical literature where the British system is used. For the record, we list other expressions in Table A3.3 such as the pound mass (lb_m) and "slug." It is useful not to get bogged down in these and other cumbersome details of the British system; clarity is assured if we stick to the pound force and its metric equivalent, the newton. Also, remember that the kilogram is a unit of mass.

Appendix Four

Suggestions for Further Reading

The footnotes found in the text refer to specific journal articles and books from which thoughts or quotations are taken. Unless the reader wants to learn further details, these sources can be ignored. Occasionally one of the authors in the following list is cited. The citation refers to this appendix, giving the author's name and sometimes the title of the work as well. The literature ranging from history and basic fluid mechanics via aerodynamics, aeronautics, and the earth's atmosphere to astronautics is immense. The following annotated list is restricted to suggestions of a few sources, most of which are readily accessible.

An introduction to basic fluid mechanics is found in many mathematical engineering texts not listed here; it is also possible to augment the material given in this text by looking up terms of interest (aerodynamic lift, boundary layer, etc.) in encyclopedias of a general or scientific nature, such as *The McGraw-Hill Encyclopedia of Science and Technology*. Many other reference books are available in every public library: among those describing fluid mechanics particularly well is Eric M. Rogers's *Physics for the Inquiring Mind* (Princeton, N.J.: Princeton University Press, 1960).

Aerodynamics and Flight

John D. Anderson, Jr. *Introduction to Flight,* third edition. New York: McGraw-Hill Book Company, 1989.

John D. Anderson, Jr. *Hypersonic and High Temperature Gas Dynamics*. New York: McGraw-Hill Book Company, 1989.

Maurice Rasmussen. *Hypersonic Flow.* New York: John Wiley, 1994.

These are engineering books. However, Anderson links the material to history, and Rasmussen provides an introduction without mathematics that is easily read.

Richard S. Shevell. *Fundamentals of Flight.* Englewood Cliffs, N.J.: Prentice-Hall, Inc., 1983.

D. Küchemann. *The Aerodynamic Design of Aircraft.* New York: Pergamon Press, 1978.

These two books deal in more detail with the airplane itself, and in the case of Küchemann—a text not readily available—with its future. Shevell gives practical tables on performance, aircraft dimensions, jet turbines, and so forth. Like Anderson, Shevell enlarges the discussion presented here of the atmosphere, the medium of flight. Both Shevell and Küchemann have many photographs of airplanes and jet turbines.

Theodore von Kàrmàn. *Aerodynamics: Selected Topics in the Light of Their Historical Development.* Ithaca, N.Y.: Cornell University Press, 1954.

In this book, one of the masters of the field, whom I have cited frequently, relates his own life experiences in conjunction with topics in aerodynamics. There is minimal use of mathematics. The material ought to be accessible to any interested reader who has plowed through this text. Subtle details of the important phenomena are described in words to deepen our understanding.

Airplanes and Spacecraft

Paul Jackson, Editor in Chief. *Jane's All the World's Aircraft.* Cohlsdon, Surrey: International Thomson Publishing Company, 1995–96.

In this large volume all that one might like to know about airplanes existing anywhere is described, including many illustrations. Jane's publishes similarly exhaustive tomes on ships, weapons, etc.

Frank Davis Adams. *Aeronautical Dictionary.* Washington, D.C.: National Aeronautics and Space Administration, U.S. Government Printing Office, 1959.

This dictionary gives a remarkably inclusive alphabetical listing of over 100,000 aeronautical terms. Moreover, fluid-mechanical entries (e.g., Bernoulli's law) are given together with entries on jet turbines, rockets, and the like. An extensive bibliography is appended. One can only hope that such unusually inexpensive Government Printing Office publications will remain in print.

Walter J. Boyne. *The Smithsonian Book of Flight*. Washington, D.C.: Smithsonian Books, and New York: Orion Books, 1987.

Boyne's book is largely a nontechnical pictorial history filled with stunning black-and-white and color photographs, drawings, and reproductions of paintings. Military aircraft of both world wars and beyond are included.

Roger E. Bilstein. *Flight in America 1900–1983*. Baltimore: The Johns Hopkins University Press, 1984.

This book is the "most comprehensive survey of the history of American aeronautics and space flight yet published" (*Technology and Culture*). In 356 pages we find an easily readable text with many black-and-white photographs and a detailed index. Military aircraft are included. Under "Notes" the basic literature for each chapter is critically discussed.

R. E. G. Davies. *Airlines of the United States*. Washington, D.C.: Smithsonian Institution Press, 1982.

Myron J. Smith, Jr. *Passenger Airliners of the United States 1926–1986. A Pictorial History.* Missoula, Mont.: Pictorial Histories Publishing Co., 1987.

Henry Ladd Smith. *The History of Commercial Aviation in the United States*. Washington, D.C.: Smithsonian Institution Press, 1991.

William Green and Gordon Swanborough. *The Illustrated Encyclopedia of the World's Commercial Aircraft*. New York: Crescent Books, 1978.

There is, of course, overlap among these books. Since public libraries are often not well stocked in this field, possibly one of these works can be found.

Charles Burnet. *Three Centuries to Concorde*. London: Mechanical Engineering Publications Limited, 1979.

Burnet's book is not readily available. It includes a nontechnical discussion of the development of the Concorde, going back to the fateful British diving experiments of the 1940s to break the "sound barrier." The book begins with a brief history of flight, with emphasis on Robins and Cayley.

William Stockton and John Noble Wilford. *Space-Liner: Report on the* Columbia *Voyage into Tomorrow.* New York: Times Books, 1981.

Richard S. Lewis. *The Voyages of Columbia, the First True Spaceship*. New York: Columbia University Press, 1984.

Milton O. Thompson. *At the Edge of Space: The X-15 Flight Program*. Washington, D.C.: Smithsonian Institution Press, 1992.

More on the advances of supersonic flight technology and the first steps into space.

Paul Garrison. *Illustrated Encyclopedia of General Aviation,* second edition. Blue Ridge Summit, Pa.: Tab Books, Inc., 1990.

This alphabetically arranged compendium of information on aviation covers aerodynamics, airports, avionics, piloting, the airplane, and much more. The clear text is accompanied by simple drawings, and the multitude of terms explained fill gaps in the book at hand.

Michael J. T. Smith. *Aircraft Noise*. Cambridge: Cambridge University Press, 1989.

Jack Williams. *The Weather Book*. New York: Random House, 1992.
Two books that augment our knowledge of flight. In nontechnical language, Smith discusses the noise produced by aircraft, and its impact on the environment. Such procedures as thrust reversal to slow down an aircraft on the ground are clearly described with many illustrations. Williams gives broad coverage in simple terms to jet streams, storms, and other aspects of weather closely linked to flight.

Airships

There still exists a retrospective fascination with the huge, rigid zeppelins that safely crossed the oceans on regularly scheduled passenger flights. There was even a small grand piano in the dining room, and all this took place long before the airplane could take over.

John Toland. *The Great Dirigibles, Their Triumphs and Disasters*. New York: Dover Publications, Inc., 1972.

A fascinating book on the great airships, dirigibles, and zeppelins. Popularly written but full of serious content. References in the footnotes of this text will lead to further information for those who admire the huge airships that regularly flew across the oceans with passengers long before airplanes could do so.

Harold G. Dick and Douglas H. Robinson. *The Golden Age of the Great Passenger Airships* Graf Zeppelin *and* Hindenburg. Washington, D.C.: Smithsonian Institution Press, 1985.

Peter Brooks. *Zeppelin*. Washington, D.C.: Smithsonian Institution Press, 1992.

History of Flight

Aspects of the history of flight are included in many of the previously listed books. It is enlightening to read Gibbs-Smith, a foremost historian of early flight whose books unfortunately are not readily found in public libraries, since most of his work was published in England. At the Science Museum in London one can buy small pamphlets by this author on many aspects of flight.

Charles H. Gibbs-Smith. *The Invention of the Aeroplane.* New York: Taplinger Publishing Co., Inc., 1965.

A carefully weighed view of the early contributions to heavier-than-air flight.

Charles Gibbs-Smith. *Early Flying Machines 1799–1909.* London: Eyre Methuen, 1975.

This is primarily a picture book, drawing largely on British and U.S. archives. An introduction relates the history.

Michael J.H. Taylor and David Mondey. *Milestones of Flight.* London: Jane's Publishing Co. Ltd., 1983.

A remarkable catalogue of aeronautical events, including flights of kites, balloons, and rockets in addition to heavier-than-air flying machines. This well-illustrated compilation clearly describes major *and* minor contributions, the title of the volume notwithstanding. An excellent index provides quick reference to happenings in aviation from 863 B.C. to June 1982 on a day-by-day basis.

Tom D. Crouch. *A Dream of Wings: Americans and the Airplane, 1875–1905.* Washington, D.C.: Smithsonian Institution Press, 1989.

A beautiful book that narrates the fascination of Americans with the development of the airplane. While it reads like a novel, the book gives the history of flight in depth. The impact of different events on the American scene is emphasized.

Tom D. Crouch. *The Bishop's Boys.* New York: W. W. Norton & Co., 1989.

More on the Wright brothers—a delightful book.

The Wright Flyer, an Engineering Perspective, ed. Howard S. Wolko. Washington, D.C.: Smithsonian Institution Press, 1987.

The Wright Brothers, Heirs of Prometheus, ed. Richard P. Hallion. Washington, D.C.: Smithsonian Institution Press, 1978.

Orville Wright. *How We Invented the Airplane,* ed. Fred C. Kelly. New York: Dover Publications, Inc., 1988.

These three inexpensive booklets give the essence of the unique achievements of the Wright brothers. The first of them describes modern wind-tunnel testing of a replica of the Wright *Flyer I* and studies of the engine, providing the basis of a full understanding of the feats accomplished by the Wright brothers. Specifically, the problems of stability—or lack of it—are investigated. I urge everyone to read the last booklet, to hear Orville Wright's voice and gain an understanding of the personalities that dominated the early period of heavier-than-air flying machines.

Tom D. Crouch. *Charles A. Lindbergh: An American Life.* Washington, D.C.: Smithsonian Institution Press, 1977.

This book leads us toward modern flight.

Michael H. Gorn. *The Universal Man: Theodore von Kármán's Life in Aeronautics.* Washington D.C.: Smithsonian Institution Press, 1992.

This book is a history of the impact of one man on education in rocketry and aerodynamics, U.S. aviation in World War II, the revival of aerodynamics in Europe after the war, and that lasting impression on many individuals, among them the author.

The Lore of Flight, revised edition, ed. John W. R. Taylor. New York: Mallard Press and BDD Promotional Books, Inc., 1990.

Appendix Five

Study Guide

Some of the problems that follow have been used in courses addressed to undergraduates in Yale College not majoring in the sciences or engineering. An instructor who chooses this book as a text or as additional reading for a similar course—or even a technical course—will probably want to add his own problems. Those who are reading this book simply because they are interested in aerodynamics may wish to test themselves. For these readers, the problems may be viewed as a survey of questions that can be answered by a close reading of the text without resorting to additional literature. The data required for the solution of the problems, such as property values, are all provided in the book.

Prior to diving into the problems that demand numerical solutions, it is useful to recall some of the material we have discussed. In no particular order, here are questions that should be answered without recourse to the text.

Write down Newton's law of force (in words if you wish). Roughly when did Newton live? Define Mach number; why is it dimensionless? Who was Ernst Mach? Describe the basic flow conditions to which Bernoulli's equation is applicable. Why does it fail for flows approaching—or even exceeding—the speed of sound? What is the speed of sound, and what is its approximate value at sea level? Recalling Bernoulli's equation, how do you explain a roof flying off in a storm, a convertible top bulging outward when a car is driven at high speed, a paint sprayer, and the crooked tube the dentist sticks in your mouth to keep it dry?

Explain the concept of the boundary layer. How would you describe laminar and turbulent flows? Think of rising smoke, clouds, and spiral nebulae. How does a hot-air balloon rise into the sky? What is a similarity parameter? Where do you apply the Reynolds number and what is its physical meaning? If you are on the moon and step on a scale operating with a spring, do you weigh more or less than at home? Otto Lilienthal and the Wright brothers played leading roles in the history of flight around the turn of the century. What did they do? If somebody were to ask you how the huge jumbo jet, the Boeing 747, manages to take off, what would you tell him? Why do your ears pop when you take an elevator to the top of the Empire State Building? Consider a small airplane and a submarine. To perform model tests on both using a wing (or water) tunnel, which primary similarity parameters would you apply? Would a surface ship obey the same law? What force ratios are expressed by the Reynolds number? What similarity parameters enter in the design of the Concorde? What makes the Concorde's aerodynamic design fundamentally different from that of a small airplane?

Now let us turn to the problems. No knowledge of calculus is required, and a simple calculator will be sufficient to perform the work. The emphasis for numerical solutions is on the SI system (see Appendix 3), a strange vocabulary to most Americans who are not scientists or engineers. Although formally adopted by the United States, the metric system is used only occasionally in this country. This fact makes us unique among the developed nations; even the British have now reluctantly parted with the inch and the yard. Occasionally, a source in the text (table or figure) is indicated, but references to the frequently needed tables for conversion of units (Appendix 3) are not repeated. The topic of dimensions—alien to those outside the physical sciences—arises often. It permits us to view the physics of a problem without an immediate involvement in such detailed activities as calculations and choice of units.

1. Give the dimensions of mechanics and their symbols. How can you use force rather than mass in calculations? Give the SI and British units of force.

2. The dimensions of energy per unit mass are given by the square of speed $(L/T)^2$. Derive this fact and discuss the result.

3. What are the dimensions of speed, acceleration, force, pressure, density, and specific weight? Use units that you are familiar with to describe terms (e.g., speed can be expressed in mph).

4. What is Newton's concept of dynamic viscosity? Discuss the implications of this idea.

5. The radius of the earth is 6,370 km. Assume that the earth is spherical. Its average density is 5.52×10^3 kg/m^3. What are the length of the equator and the volume of the earth and its mass in the SI system of units? (Needed formulas are given in Appendix 1.)

6. A filled water tank has a volume of $4 \times 6 \times 12$ m^3. Find the weight of the water in the tank in newtons and lbs. Use powers

of ten. (See Table 3.1 for the density of water and Table 4.1 for acceleration of gravity. Remember that specific weight equals density times acceleration of gravity.)

7. Write the following numbers in powers of ten (e.g., $100 = 10^2$): 1,000,000; 0.01; 0.04; 1; 1/1,000.

8. Assume that your specific weight is equal to that of water. (Is that a good assumption, considering your experience when swimming?) Compute your volume based on your weight and express it in m^3 and ft^3. Explain your answer. (Use Tables 3.1 and 4.1 to compute the specific weight of water from its density.)

9. Would a hydrogen or a helium balloon of a given size be able to lift more weight into the air?

10. Iron ships were first built in the last century. How would you have convinced a frightened sailor that they wouldn't sink?

11. Why is it easier to swim in the ocean than in fresh water?

12. From a spring balance calibrated in newtons is hung a mass of 1 kg. What reading do you get in newtons? Can you determine the acceleration of gravity (Table 4.1) from this experiment? Explain your answer. If an astronaut takes the same balance and mass to the moon or Mars, could he determine the acceleration of gravity of these bodies carrying out the same experiment?

13. A rock weighs 200 lbs on a spring scale, and when it is lowered into a cylindrical tank of water with a 2-ft diameter, its weight is 137.7 lbs. How much does the water rise in the tank? (The specific weight of water is about 62 lbs/ft^3.)

14. A female astronaut weighs 110 lbs on earth, and she can readily carry a pack weighing 20 lbs. What is the approximate *earth* weight of the equipment that she can carry on Mars without exceeding the muscular effort needed on earth to carry herself and the 20 lbs? (The acceleration of gravity on Mars is 3.74 m/s^2.)

15. In the lower atmosphere the lapse rate is found to be about –6.5°C/km. If the temperature at sea level is 20°C, what will it be on a mountain 5,000 ft high in °C and °F?

16. An aircraft flies at an altitude of 12 km. The cabin is pressurized to the equivalent of an altitude of 2 km. A window on the plane has an area of 1 ft^2. Compute the force exerted on the window in lbs and N and give its direction.

17. A swimming pool located at sea level has a depth of 3 m. What is the pressure at the bottom of the pool, in units of your choice? Remember, there is atmospheric pressure at the top (Table 4.1).

18. In Chapter 5, you find Bernoulli's equation in the form

$$p_o = p + (\rho/2)\, u^2.$$

What are the assumptions implicit in this version of the law of conservation of energy? Name the three terms, give their dimensions, and derive an expression for the speed u. How do you measure the speed?

19. A duct of varying cross section has a discharge rate of $Q = 100$ m^3/s in steady flow. What is the flow speed (one-dimensional) at cross sections of 1 m^2 and 5 m^2? How do you get the answers?

20. An aircraft flies at an altitude of 2 km at a speed of 200 mph. What is the stagnation pressure, p_O, in pascals (the SI unit of pressure, N/m^2), and in psia? With what instrument is this pressure measured, and where would you locate it on the airplane? Consult Table 4.1 and Appendix 3.

21. Two boats are side by side at anchor on a river, a short distance from each other. If the speed at which water flows in the river increases, what happens to the distance between the boats? Explain your answer.

22. Why can the roof of a building fly off in a high wind? Is there anything you can do prior to a storm to prevent this from occurring?

23. A car with a cross section $A = 3$ m^2 is driven at 30 mph. Its drag is $D = 70$ lbs.

    a) What is the drag coefficient of the car? Is it an aerodynamically good car?

    b) With c_D = constant, what is the air drag at 60 mph?

    c) What is the required power at 30 and 60 mph in kW? (Ignore rolling and internal friction.)

    d) What is the average drag coefficient of current cars?

24. A disk and a streamlined body have drag coefficients of $c_D = 2$ and 0.05 respectively. What is the ratio of the cross-sectional areas of the two bodies, provided the *drag* of both is equal at equal speed and air density?

25. A baseball is thrown at sea level at 100 mph. (This is about the best a major-league pitcher can do!)

    a) Find the size of a standard baseball and compute its Reynolds number based on diameter, using Table 3.1. Assume the baseball to be a perfect sphere, and find its drag coefficient in

Figure 7.4. (Watch the logarithmic scale!) Because the baseball has seams, you should assume that the boundary layer is turbulent. Why?

b) What is the air drag of the ball in lbs?

c) How long does the ball take to travel from the hand of the pitcher to the bat, assuming that its speed remains constant during flight?

d) In light of the Magnus effect, how would you explain a curve ball?

e) Would the ball move more slowly if it had a laminar boundary layer? Explain.

26. A ship has a length at the waterline of 120 m and moves at 30 mph. What is the Reynolds number? (See Table 3.1 for properties.) Do you expect a largely laminar or a turbulent boundary layer on the ship below the water surface (see Figure 7.7)?

27. The flight conditions of a certain airplane are characterized by a Reynolds number of 10^6. From the equations for the friction coefficients of laminar and turbulent boundary layers on a flat plate given in Chapter 7, calculate the two values. Check your calculation by looking at Figure 7.7. If you assume that wall friction on a wing is about equal to that on a flat plate, you can apply your calculation to the aircraft. Now determine the ratio of the actual friction forces of the two types of flow. Discuss this ratio in light of Figure 10.5 and the idea of a "laminar wing." (Figure 10.5 applies to a current aircraft with a turbulent boundary layer on the wing.)

28. An airfoil model has a drag of D = 2 lbs and a lift of L = 24 lbs. Give the approximate value and direction (with respect to the flight path) of the resultant air force. You may solve this problem graphically.

29. A housefly has a wingspan of 0.013 m, and its optimum flying speed is 2 m/s. What is its Reynolds number at sea-level flight, based on wingspan? (Use Table 3.1 for properties.)

30. An airplane flies in the stratosphere at a cruising speed of M = 0.8. The speed of sound in the stratosphere is 295 m/s.

a) What is the cruising speed in km/h and mph?

b) The aircraft takes 5 hours to cover a distance of 3,000 miles. What is the average flight speed for the trip?

31. What similarity parameters are required to test a model of a supersonic aircraft? What is physically implied by the similarity parameters?

32. The speed of sound at sea level is about 760 mph. It is proportional to the square root of the absolute temperature designated by K. An airliner flies at $M = 0.8$ at 10-km altitude; what is its ground speed? (For the definition of absolute temperature and temperature as a function of altitude, see Table 4.1.)

Here are solutions for some of the problems.

1. Study Appendix 3 and write down each term to fully understand it. Add various units of your choice.

2. Again the answers can be found in Appendix 3. However, the facts derived suggest an interesting diversion. Einstein's famous equation $E = mc^2$ of the special theory of relativity—often quoted without explanation in the popular press—can now be better understood. Here E is the energy, m the mass, and c the velocity of light. The latter is 300,000 km/s or 186,000 miles/s. Einstein's equation reveals the equivalence of mass and energy. Since the velocity of light appears squared in the expression, we deduce that even a very small amount of mass is equivalent to an enormous amount of energy. Write dimensions and units of your choice next to the terms of Einstein's equation to understand it better.

5. Slicing the earth in two (equal) halves, you find the circumference of the cut, c, to be the equator. From $c = 2\pi r$ with $\pi = 3.14$ and the radius of the earth, you calculate c = 40,000 km. In the last century when the metric system was introduced, the French wanted to tie the meter to a global dimension. They defined it as the forty-millionth part of the equator, a number that could not be measured accurately at the time. The volume of the spherical $V = (4/3)\pi r^3 = 1.1 \times 10^{21}$ m^3. Multiplying the volume by the value of the density of the earth, the mass is 6×10^{24} kg, a number too large to be perceived. The earth is, of course, not a ball of uniform density. Drilling a hole from the surface to its center, we actually encounter a thin crust, molten materials, and a liquid metallic center, all of different densities. Our value of density represents a mean value of all the different materials of our planet.

6. With the tank volume $V = 288$ m^3, density of water (Table 3.1), and acceleration of gravity (Table 4.1), the weight of the water, W, is given by $W = V\rho g = 2.8 \times 10^6$N. Using a conversion table (Appendix 3), $W = 6.3 \times 10^5$ lbs.

8, 9, 10, 11. All four problems rely on Archimedes' principle of buoyancy discussed in Chapter 3.2. For problem 9 you need the atomic weights of hydrogen, 1, and helium, 4, to tell you which is the lighter gas.

12. A newton is defined as the force which gives a mass of 1 kg the acceleration of 1 m/s^2. What is the weight of 1 kg mass in newtons determined with a spring balance on earth (for g see Table 4.1)? Mars and the moon have accelerations of gravity of 3.7 m/s^2 and 1.6 m/s^2, respectively. From the definition of N you note that the numerical value of the acceleration of gravity (not its dimension!) is equal to that of the weight of the kg mass in newtons. Consequently, the mass of one kg weighs 9.8, 3.7, and 1.6 N on the three heavenly bodies. In turn, equipped with a spring balance and a mass of one kilogram, you can determine acceleration of gravity at any location.

14. The astronaut with backpack weighs 130 lbs on earth. Assume that her muscle power is unchanged on Mars. With $g = 3.7$ m/s^2 on that planet, we find $g_{Mars}/g_{Earth} = 0.38$. Therefore on Mars using her spring scale, she finds she weighs only 42 lbs without her pack. Since we know from earth that she can handle 130 lbs, she can carry another 88 lbs on Mars, since 42 + 88 = 130 lbs. But a weight of 88 lbs on Mars will be 232 lbs on earth, permitting her to handle a huge pack on Mars.

15. 5,000 ft = 1,530 m = 1.53 km. With –6.5°C/km, we find a cooling of the air by about 10°C. Starting at 20°C, the temperature at 5,000 ft will be 10°C or 50°F. (It is easy to remember that 10°C = 50°F.)

16. Table 4.1 shows that pressure at 12-km altitude is given by $(p/p_0)p_0 = 0.192 \times 14.7$ psia = 2.82 psia. (Remember that psia denotes the absolute pressure counted from zero). The cabin pressure is equal to that of the atmosphere at 2-km altitude, a value that insures comfort. We find by the above method that $p = 11.5$ psia inside the cabin. Therefore the pressure difference 11.5 – 2.82 = 8.68 psi is acting on the window, and in fact on all cabin walls. In turn, one square foot (144 square inches) has to sustain a force of 1,250 lbs = 5,560 N. Obviously the force acts from the inside out; the cabin is a huge pressure vessel.

18. For the airspeed measurement with a pitot-static tube (Figure 5.7), you rewrite the Bernoulli equation in terms of speed. Remember that ρ = constant in incompressible flow. The result is $u = [(p_0 - p)/(\rho/2)]^{1/2}$, where the power of 1/2 indicates that we take the square root of the equation in brackets.

19. $Q = uA = 100$ m^3/s; $u = Q/A$. For $A = 1$ m^2, $u = 100$ m/s; for $A = 5$ m^2, $u = 20$ m/s. Think of the general function of speed as a function of the cross section of the duct (Chapter 5.2).

20. Another Bernoulli problem for which you find the equation expressing the stagnation pressure in Chapter 5 or in your work on problem 19. Table 4.1 tells you that at h = 2 km we have for the atmospheric pressure $p = (p/p_0)p_0 = 0.785$, $p_0 = 7.95 \times 10^4$ Pa. Here the unfortunate SI pressure unit, the pascal, turns up with Pa = N/m^2. The atmospheric density is also lower at 2-km altitude than at sea level. From Table 4.1 we calculate $\rho = (\rho/\rho_0)\rho_0 = 0.822\rho_0 = 1.01$ kg/m^3. The flight speed of 200 mph = 322 km/h = 89 m/s. Inserting these values in Bernoulli's equation, we find $p_0 = 8.8 \times 10^4$ Pa = 0.86 atm = 13 psia. The pitot tube, an open pipe facing the wind, is mounted on the nose or the wing of the aircraft, as you may have noted. I have stressed the pascal to prepare you for weather reports that you may hear in Europe. There the barometric pressure is mostly given in this unit in parallel with temperature in Celsius.

21. Picture the two boats from above and view Figure 5.5. Between the boats the river speeds up and the boats will move closer.

22. It is useful to open attic windows to equalize the indoor and outdoor pressures. In a storm the pressure on the roof drops (Bernoulli), and the inside pressure of the house can lift the roof.

23. a) $c_D = D/(Aq)$ with the dynamic pressure $q = (\rho/2)u^2$. With D = 70 lbs = 312 N, and the dynamic pressure $q = 10^4$ in pascals, we find $c_D = 311/(3 \times 111) = 0.94$. A high drag coefficient, like that found for a moving box (Figure 7.10).

    b) Doubling speed from 30 to 60 mph increases the drag four times to the value 280 lbs.

    c) Power equals drag times speed. For our car we find 4 kW (kilowatts) at 30 mph and 32 kW at 60 mph. Table A3.3 permits you to convert these power values to horsepower, a more familiar unit. In Europe, the engine power of cars is often indicated in kilowatts.

    d) A drag coefficient of about 0.3 would be good for a car; most still have higher values.

25. I leave this problem to the reader, since most concepts have appeared in previous problems. Remember that $\nu = \mu/\rho$ is a constant in incompressible flow at a given temperature (Table 3.1).

26. Calculate the Reynolds number of the ship with the properties of water also found in our much-used Table 3.1. With length = 120 m, u = 30 mph = 13.4 m/s, and $\nu = 10^{-6}$ m^2/s, we find the Reynolds number Re = 1.6×10^9. Check dimensions and units. Assuming the hull acts aerodynamically like a flat plate, the boundary layer will clearly be turbulent, as seen in Figure 7.7.

27. For $Re = 10^6$, we find from the equations of Chapter 7 friction coefficients of 1.3×10^{-3} and 4.7×10^{-3} for laminar and turbulent flow, respectively. The turbulent boundary-layer friction is close to four times as high as that of the laminar condition. At higher Reynolds numbers of jet transports (see Figure 6.3), the difference of the friction between the two types of boundary layers becomes even more pronounced, indicating how desirable a laminar wing would be (Figure 11.1).

28. Look at Figure 1.1 and draw a lift-and-drag rectangle just like that on top of the aircraft. For example, choosing a scale of one inch to the pound, you can measure the length of the resultant air force immediately in inches to find its force in lbs. The required force and angle with respect to the flight path can also be found from trigonometry.

29. The answer to problem 30 can be found as before.

30. a) Recall the definition of the Mach number from Chapter 6.1. We have $u = M \times a = 0.8 \times 295$ m/s = 236 m/s = 850 km/h = 528 mph.

    b) Too easy to provide a solution!

31. Models of a supersonic airliner ought to be tested at the Mach number *and* the Reynolds number of the full-size airplane. Mach number can readily be simulated in modern wind tunnels. To simultaneously achieve the high required Reynolds numbers in the same test is not yet possible. However, by calculations based on the extensive experience at lower Reynolds numbers in wind tunnels and flight tests, this problem has been overcome, as seen in the peerless performance of the Concorde.

32. Find the temperature at the altitude of flight in degrees Celsius (Table 4.1) and convert it to degrees Kelvin by the definition of the absolute temperature. The square root of the ratio of the absolute temperatures at flight altitude and ground level, $(223/288)^{1/2} = 0.88$, provides the temperature effect on the speed of sound. This term multiplied by the speed of sound at sea level gives the speed of sound at the flight altitude, $0.88 \times 760 = 669$ mph. Since the airplane flies at $M = 0.8$ (much like a real airliner), we have—as before—for the ground speed $u = M \times a = 0.8 \times 690 = 535$ mph (compare Table 11.1).

Figure and Table Credits

Figures

Deutsches Museum, Munich
1.2, 1.4, 1.11, 1.12, 2.2, 2.4, 2.5, 10.9, 11.4

Charles Gibbs-Smith, *Early Flying Machines 1799–1909,* Eyre Methuen, 1975
1.5, 1.8, 1.10, 1.15

Science Museum, London
1.6 (neg. 291/55 and 292/55), 1.9 neg. SM312

Charles H. Gibbs-Smith, *Sir George Cayley's Aeronautics 1796–1855,* Her Majesty's Stationery Office, 1962
1.7

John D. Anderson, Jr., *Introduction to Flight,* third ed., © 1989, p. 21. Reproduced with permission of McGraw-Hill, Inc.
1.13

National Air and Space Museum, Smithsonian Institution, Washington, D.C.
1.14, 2.3, 2.6

Illustration from Henry T. Wallhauser, *Pioneers of Flight,* Hammond Incorporated, Maplewood, N.J., 1969
2.1

Photo courtesy of Paul MacCready, Monrovia, California
2.7

Photo courtesy of Thomas J. Mueller, University of Notre Dame
5.3, 8.5

Peter P. Wegener, Yale University
5.6, 10.2, 10.7

Guggenheim Aeronautical Laboratory, California Institute of Technology
6.1, 7.11, 11.4

National Aeronautics and Space Administration
6.2, 10.10

John H. McMasters, *American Scientist* 77:167 (1989). Reprinted by permission of *American Scientist,* journal of Sigma Xi, The Scientific Research Society.
6.3

Sighard F. Hoerner, *Aerodynamic Drag,* revised ed., privately published, Midland Park, N.J., 1965
7.3, 7.10

Richard S. Shevell, *Fundamentals of Flight,* second ed., copyright 1989, pp. 49, 242. Reprinted by permission of Prentice Hall, Inc., Englewood Cliffs, N.J.
8.6, A2.3

Prandtl-Archive, Göttingen, Germany
8.12

S. Goldstein, ed., *Modern Developments in Fluid Dynamics,* vol. 2, Clarendon Press, 1938
8.14

National Agricultural Library, U.S. Forestry Service Photo Collection
8.15

W. Nachtigall, University of Saarbrücken, Germany
8.18

Richard S. Shevell, Technological development of transport aircraft—past and future, *J. Aircraft* 17:67–80 (1980). Reprinted with permission.
10.1

R. Courant and K.O. Friedrichs, *Supersonic Flow and Shock Waves,* Interscience Publishers, 1948. Reprinted by permission of Mr. Ernest Courant.
10.3

Transonic Wind Tunnel, Göttingen, Germany, courtesy of Deutsche Forschungsanstalt für Luft- und Raumfahrt, e.V., Wind Tunnel Department
10.4

Peter P. Wegener, The science of flight, *American Scientist,* 74:268–278 (1986). Reprinted by permission of *American Scientist,* journal of Sigma Xi, The Scientific Research Society.
11.1

Reprinted with permission from *Cities and Their Vital Systems,* © 1988 by the National Academy of Sciences. Published by National Academy Press, Washington, D.C.
11.2

Ronald L. Bengelink, Boeing Commercial Airplane Group: R. Hilbig and J. Szodruch, MBB, AIAA paper 89–0534 (1989). Reprinted with permission.
11.5

Tables

U.S. Standard Atmosphere, U.S. Government Printing Office, Washington, D.C., 1963
4.1

Sighard F. Hoerner, *Aerodynamic Drag,* revised edition
7.1

Data from: John H. McMasters, Boeing Commercial Airplane Company; Reinhard Hilbig, Messerschmitt-Bölkow-Blohm, Transport Aircraft Group; Richard S. Shevell, Stanford University; Ronald L. Bengelink, Boeing Commercial Airplane Group; Myron J. Smith, Jr., *Passenger Airliners of the United States 1926–1986,* Pictorial Histories Publishing Co., 1987
11.1

Name Index

Entries with page numbers followed by *n* appear in the footnote of that page.

Subject Index

Entries with page numbers followed by *n* appear in the footnote of that page.